Integrated Occupational Safety and Health Management:
Solutions and Industrial Cases

职业安全与健康综合管理
——解决办法与工业案例
（影印版）

〔芬〕Seppo Väyrynen，Kari Häkkinen，Toivo Niskanen

程健维　中文导读

科学出版社

北　京

图字：01-2020-0494 号

内 容 简 介

本书系为安全工程专业实施全英文教学而引进的，内容符合"职业健康与安全"课程教学大纲的要求，由来自芬兰的资深安全专家共同编写，采用理论与案例结合的方式进行组织内容是对职业健康和安全进行综合论述的畅销书。全书共15章，主要内容包括：绪论、安全管理、建筑行业事故原因及预防、事故信息在安全促进中的应用、HSEQ评估与应用、港口HSEQ案例研究、驾驶室外卡车司机安全研究、危险液体运输案例、典型用人单位事故损失风险分析、HSEQ培训方法、化学工业工作场所化学品安全、领导力和职业安全、企业社会责任、企业可持续发展研究等。

本书可供安全工程、环境工程及公共与预防医学、应急管理专业的本科生、研究生及相关工程技术人员进行学习与参考。

First published in English under the title
Integrated Occupational Safety and Health Management: Solutions and Industrial Cases
edited by Seppo Väyrynen, Kari Häkkinen and Toivo Niskanen, edition: 1
Copyright © Springer International Publishing Switzerland, 2015 *
This edition has been reprinted and published under licence from
Springer Nature Switzerland AG., part of Springer Nature.
For copyright reasons this edition is not for sale outside of China Mainland.

图书在版编目（CIP）数据

职业安全与健康综合管理：解决办法与工业案例 = Integrated Occupational Safety and Health Management: Solutuions and Industrial Cases：英文/(芬) 塞波·罗特 (Seppo Vayrynen) 等著. —影印本. —北京：科学出版社，2020.3

ISBN 978-7-03-064682-8

Ⅰ. ①职⋯ Ⅱ. ①塞⋯ Ⅲ. ①劳动保护-劳动管理-英文 ②劳动卫生-卫生管理-英文 Ⅳ. ①X92②R13

中国版本图书馆 CIP 数据核字（2020）第 041108 号

责任编辑：李涪汁/责任校对：郑金红
责任印制：张 伟/封面设计：许 瑞

科学出版社 出版
北京东黄城根北街 16 号
邮政编码：100717
http://www.sciencep.com

北京捷迅佳彩印刷有限公司 印刷
科学出版社发行　各地新华书店经销
*

2020 年 3 月第 一 版　　开本：720×1000　1/16
2020 年 3 月第一次印刷　　印张：21
字数：420 000

定价：159.00 元
（如有印装质量问题，我社负责调换）

前　言

这本书的"故事"起源于健康、安全、环境、质量（health, safety, environment, quality）的首字母缩写 HSEQ。2012 年春季，编者和 Springer 出版社就 HSEQ 的关键内容通过邮件交换了看法。随后，新组成的编者小组安排了一个为期一天的集中研讨会，以便通过更有条理的方式进一步开展本书的编写。很快，编者们就以思维导图的形式将本书的第一个关键想法勾勒出来了（图 1）。

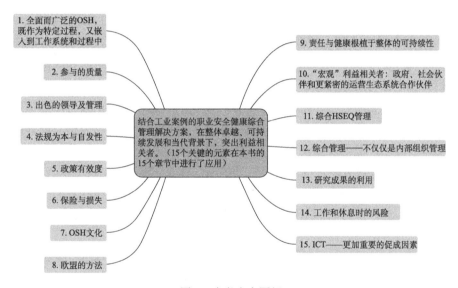

图 1　本书内容图解

本书各章的写作过程均按照计划进行，同时牢记思维导图中所展示的当代职业健康领域的发展趋势及前提，所有这些都是本书编著者，也是这一领域的专家和从业人员典型的基本工作方法。

图 1 包含的重点以一个更加宏观而凝练的方式展开：

（1）职业安全和健康管理（occupational safety and health management, OSHM）是一个集合了多学科专业知识和管理的特别领域。同时，它必须"有机"地融入每个工作场所的所有管理工作。

（2）组织内有利于提高整体质量的所有资源都是必需的。为了达到这一目的，个人及组织层面都需要广泛而透明的参与。

（3）当代卓越的管理强烈促进了 OSHM 的发展。

（4）对 OSHM 的评价基于通用立法和组织特定的驱动因素，以提高责任竞争力和公众形象。

（5）当局行动和监管职能试图寻找一种协同作用来拓宽 OSHM 渠道，并进一步发现政策有效性的证据。

（6）保险业是联系和支持所有工作组织的重要机构，涉及所有的风险、损失和资产管理。

（7）一个组织努力实现自己的 OSHM 文化，并将其作为一项基本要素嵌入自身文化中。

（8）欧盟主流模式受到北欧国家以往应对挑战风格的强烈影响。

（9）社会责任、可持续性、工作中的福利、生产力以及商业、劳动力和环境等全球大环境作为一组驱动力，鼓励甚至强迫工作组织和关键利益相关者进行沟通和协作。

（10）在所有工作组织中，利益相关者的关键作用是由政府、雇主协会和工人工会所组成的三重角色共同发挥。管理的手段被利益相关群体所支持和鼓励，前面所提及的三个实体都是该利益相关群体的成员。

（11）综合风险管理是一个重要的概念体系，涉及 OSH 问题、具备良好实践的参与者以及每个公民所期望的目标。与这个体系相联系的一个关键因素，即广泛综合的 HSEQ 管理系统，它是应对未来挑战的答案和实用工具。

（12）不仅供应流程和"纯"业务是"新"OSHM 的关键，组织对网络价值的认识及其在开发共享工作系统中的影响也日渐重要。

（13）定量与定性辅助研究手段的开发与创新为工作组织提供了一个日益重要的机会和一个利益相关方。

（14）健康、安全和幸福（在工作中也是如此）需要每个雇员自己以及社会全天候照顾和维护。

（15）所有上述各点的一个关键促成因素是信息和通信技术（information and communication technologies，ICTs）的广泛使用，通过台式机、笔记本电脑和移动设备人们进行交流互动。在 OSHM 和 HSEQ 中，数字化正日益成为一种必要模式。

本书作者来自大学、研究机构、保险业和公共监管机构、工作组织、公司和外部利益相关方。此外，本书撰写以多种方式保证了与工厂的密切的合作。作为编者，我们衷心感谢所有撰稿人。我们也要感谢众多为本书做出贡献的人、参与其中的组织及其管理者、专家、员工、企业家、合作伙伴、研究与开发（R&D）项目组织和主要利益相关方。最后，我们感谢玛丽亚·林德霍尔姆夫人在成书的最后阶段对手稿的协助处理。

编者很希望这本书的"故事"能被读者继续下去。本书旨在为业界和学术界的读者提供新的观点和工具，以推动 OSHM 的发展。

Seppo Väyrynen

Kari Häkkinen

Toivo Niskanen

目 录

前言
第1章 绪论 ·· C-1

第一部分 发展安全管理及领导力来预防OSH损失

第2章 安全管理——从基本认识走向卓越 ·· C-4
第3章 建筑行业事故原因及预防——芬兰一些案例进展 ······················ C-5
第4章 事故信息在安全促进中的应用——工业维修案例 ······················ C-5

第二部分 职业安全与健康管理的观点及持续改进的案例

第5章 芬兰工业网络内的综合管理：钢铁厂HSEQ评估程序案例 ··· C-6
第6章 港口场景的HSEQ框架介绍：综述和芬兰北部案例 ············ C-7
第7章 驾驶室外卡车司机工作系统——宏观人机工程学发展方法 ··· C-7
第8章 危险液体运输中的致命非驾驶事故：芬兰工业系统中严重事故的调查程序及相关案例 ·· C-8
第9章 芬兰两大用人企业员工失时事故总风险 ································ C-8
第10章 芬兰北部的HSEQ培训园区——建筑业创新合作论坛 ········ C-9

第三部分 OSHM效应

第11章 芬兰化学工业工作场所化学品的安全使用与风险防范 ······ C-10
第12章 芬兰化学工业的领导力和OSH过程 ································· C-11
第13章 不同利益相关者对企业社会责任的探讨 ····························· C-11
第14章 芬兰企业可持续发展的案例研究 ······································ C-12

第1章 绪 论

如果我们要找出一种理想的当代职业安全和健康管理（OSHM）方法，可以用立法、标准和良好实践这三个基石来建立它的模型。当然，第三种是以某种方式强制实行的，也包括许多非强制性的做法，后者的驱动力通常是与 OSHM 相关的增强组织竞争力的措施。OSHM 贡献的竞争力可以产生内部或外部成果。例如，内部成果包括工作中的幸福感和生产力的提高。外部成果可以通过改善和提升客户、股东、利益相关者和公民对组织印象而达到。无论是"内部"还是"外部"，符合整体质量是一个至关重要的目标。这本书主要涵盖上述内容所涉及的各种最佳做法。

如前所述，本书作者涉及工业公司、保险行业、大学、研究机构和公共监管机构。他们提供了不同的观点，这些观点的共同特点是与相关实践、案例、多种解决办法、经验教训和风险预防程序有密切且直接的联系。后述 13 个章节中涵盖了许多不同的工业行业分支。

近几十年来，OSHM 在组织方面具有核心作用的观念一直流行。本书通过提供一个框架，来帮助管理人员在解决 OSH 问题时，能够理解 OSHM 体系，从而使 OSHM 的应用更加精确（不仅仅是核心作用概念）。在试图了解组织是否可能成功实施某种前瞻性的 OSH 措施时，管理者往往能学到很多 OSHM 中的技能。资源是促进 OSHM 的重要因素，与 OSHM 流程和价值相比，OSHM 资源通常可以更容易地跨组织边界转移。毫无疑问，获得高质量的 OSH 资源可提高组织有效且具有前瞻性地实施职业安全措施的能力。当最高管理层评估其组织是否能够成功地实施 OSH 措施时，OSH 资源是管理人员最根本的考量对象。

本书提出了一个挑战，即企业组织不必在外部压力（例如，立法和标准）的驱使下采取措施来增强自己的 HSEQ 实践。需要强调的一点是，组织、领导、部门经理和雇员之间的最佳实践联系不仅仅是行为守则（例如，在 HSEQ 管理中）。特别地，最佳的实践做法包括一系列 HSEQ 事务的沟通以及在 HSEQ 方面促进和提高企业的价值观。例如，这些问题涉及 HSEQ 人员的一些相关活动，包括业务生产、OSH 氛围、培训、监测系统、沟通渠道、社会技术工作设计和信息系统等。HSEQ 中组织策略及政策的选择对其价值观和行为方式有重要影响,然后形成"这里的做事方式"。这些要素中有许多 HSEQ 的组织文化。

多种工业分支涵盖在本书 13 个章节中。

第 2 章总结了工业企业对安全管理认识的发展,重点介绍了工业企业安全管理的特点。虽然良好的安全管理要素是众所周知的,并且写在了教科书和标准中,但安全管理的实际结果往往不是很好。这些经验表明,需要将全面管理作为安全长期发展的基础,而不仅仅是使用简单的技巧实施解决方案。

第 3 章综述了建筑工业中常见的健康和安全危害以及事故风险,包括一些事故统计数据和一起致命事故的案例描述。近年来芬兰建筑业及其利益相关方注重提高安全性所取得的成果也罗列其中。

第 4 章总结了工业维修中事故来源和风险评估的一些基本观点。重点研究了芬兰工业维修期间发生的严重和致命事故。此外,还回顾了事故保险机构的数据库,以了解工业维修中较轻的事故。典型的风险因素有系统可维护性差、任务安全规划不完善以及与客户合作的不足等。

第 5 章描述了在一般工业环境下,HSEQ 供应公司在案例公司公共工作场所中的 HSEQ 管理实践,案例公司是芬兰北部的一家钢厂。本章对供应商的 HSEQ 评估程序(HSEQ assessment procedure, HSEQ AP)进行了说明,并对 HSEQ AP 的结果以及其他安全指标进行了分析。截至 2014 年秋季,HSEQ AP 集群已经达到了相当大的规模。HSEQ AP 集群的大型采购公司多达 7 家,并且评估了约 120 家公司。

第 6 章讨论了在管理芬兰海港时使用 HSEQ AP 的必要性和实际应用的可能性。该港口位于巴伦支海地区北部。本章的编写主要是基于文献以及访谈。本章研究的实证访谈部分主要是在芬兰北部进行的,访谈的内容较为详细。在访问俄罗斯摩尔曼斯克商业海港期间,也收集了关于 HSEQ 的信息。

第 7 章讨论了本地和短途(local and short haul, L/SH)卡车司机在各种工作环境中(除了卡车驾驶室)的安全问题,这些安全问题之间的差异很大。这项研究将事故统计分析(过去)、视频观察(现在)和情景研讨(未来)三个时间角度分析方法结合起来,为交通运输行业的设计和管理过程的发展提供新的方法。

第 8 章回顾了芬兰对致命事故的调查程序,并分析了与卡车运输危险液体有关的 4 个事故案例，特别是卸货后服务操作相关的事故。运用内容分析法对调查报告中的事故原因进行了识别，为今后提供预防措施。

第 9 章针对芬兰两大工作场所雇员（$n=13000$）涉及的所有伤害性事故（10 年内）总结了最新的综合数据。本章研究旨在阐明和评估员工不同的失时工伤（LTI）事故风险，即在工作、家庭和闲暇时间以及上下班途中发生的事故。

第 10 章指出建筑业迫切需要一种新的安全培训。这种培训需要新的方法和程序，才能有效地减少芬兰目前较高的事故发生率。HSEQ 培训园区作为安全培训创新的基地建立一个全面的行业实际工作情况的模型，通过实际操作演练和主动参与来实现有效学习。

第 11 章探讨了 HES（健康、环境和安全）管理人员和 OSH 员工代表如何看待以下议题：①化学品的安全使用；②预防优先级和管理措施；③在系统思维中 OSH 的集体氛围；④关于化学品安全立法的论述。49 名 HES 经理和 105 名 OSH 员工代表参与了在线问卷调查。

第 12 章探讨了 OSH 的组织和技术措施，并阐明立法、领导力、协作、预防、改进、监测、职业卫生保健、培训和个人防护设备使用之间的潜在关系。参与者是来自化学公司的 OSH 管理人员（$N=85$）和 OSH 员工代表（$N=120$）。

第 13 章探讨了不同利益相关者如何看待企业的社会责任。主要研究的问题是：企业社会责任如何在商业活动中达到最佳实践？子问题是：企业社会责任如何在公司的商业战略、金融家和股东、客户和消费者、公司员工以及社区和政府的日常活动中达到最佳实践？

第 14 章探讨了芬兰公司在企业可持续发展方面的各种实践，企业可持续性如何应用于其商业实践。组织的价值在于为管理者和员工做出最优决策提供标准。

本书整体着重于实践，同时也考虑立法和标准。在这方面，我们要提到，国际标准化组织（ISO）/CD 45001 下的 OSHM 通用国际标准化准则将于 2016 年出版。希望即将推出的标准能够使 OSHM 应用更广泛，并提高 OSHM 在工作组织中的重要性。本书不仅限于法规和标准，还为工作组织中的人员和管理提供了相关案例。如何在实践中通过强调社会、经济和环境的可持续性以及负责任的管理来管理组织的问题在许多方面得到了回答。

第一部分 发展安全管理及领导力来预防 OSH 损失

第 2 章 安全管理——从基本认识走向卓越

摘要 本章综述了工业企业对安全管理认识的进展,重点介绍了安全管理良好的工业企业的最普遍特征。本章的基础是工业意外保险实践的知识和经验,以及选定的健康和安全方面相关文献的调查结果。成功的要素包括:良好的整体管理,安全目标,为管理者定义的能力要求,最高管理层的可见性和承诺,对安全工作的激励,有效地从事故中学习并使用风险评估持续改进的能力,有效的安全检查和内部审计,以及包括沟通和信任的参与型领导实践。这些经验表明,需要以整体管理为基础,实现安全卓越的长期发展,而不是实施单一的技巧和解决手段。近年来,高度发达的经济体建立了安全管理体系,从而改进了正式的安全管理方法。今后,需要更好地处理安全管理中非正式进程,例如人的行为、态度和安全文化。此外,健康和安全的复杂性要求改进一般管理理论,以便更好地应对管理中以人为中心的复杂问题。

第3章 建筑行业事故原因及预防——芬兰一些案例进展

摘要 即使在发达经济体中,建筑业也被认为是工伤事故风险最高的行业之一。本章综述了建筑工程中的健康、安全危害以及事故的风险。根据芬兰工人赔偿制度的索赔数据,结合一起致命事故的案例,给出了一些事故统计数据。这起致命事故表明了在几个分包商共同工作的场所高空作业的风险。人们对安全的关注促进了安全工作的积极发展,这既提高了索赔统计数字,也带来了一些改善建筑工地安全措施的创新。一些大型国际承包商在建设项目的安全管理方面起到主导作用。这些改进似乎与高层管理角色的强化密切相关,因此,现在主管和员工都更加重视安全问题。新的工具和程序都有利于更有效和更系统地实施现场安全措施,例如事件调查和报告以及工作规划程序。然而,在建筑业实现零事故文化的前进之路上仍然需要持续的努力。通过这种努力,结果才可能会有显著的改善。这意味着,有关各方的认识是一个持续不断的过程,在这个过程中应该避免自满。

第4章 事故信息在安全促进中的应用——工业维修案例

摘要 维修作业基本上是所有工业环境中都会进行的。根据操作环境、条件以及所涉及的任务的不同,风险差别很大。本章总结了工业维修领域的一些研究成果。这些研究是在2004~2014年期间进行的。主要素材从各种事故数据库中收集,并得到了公司风险评估的支持。一项主要研究集中在重新分析芬兰工业维修期间发生的致命事故和严重的非致命事故。此外,芬兰事故保险机构的数据库也被修订,以便了解工业维修中发生的较轻的事故。典型的风险因素有系统可维护性差、任务安全规划不完善和与客户合作不足等。风险评估和管理应注意不同的任务、操作环境以及任务安全规划。本文总结了工业维修事故来源及风险评估的一些基本观点。

第二部分 职业安全与健康管理的观点及持续改进的案例

第5章 芬兰工业网络内的综合管理：钢铁厂HSEQ评估程序案例

摘要 本章描述了一般工业背景下，HSEQ 供应公司在案例公司公共工作场所中的 HSEQ 管理实践。该案例公司是芬兰北部的一家钢铁厂。另外，在这个公共的工作场所经营的三个 HSEQ 供应公司也被选择作为研究对象。HSEQ AP 是芬兰七家主要供应商公司使用的管理程序。本章的目的是详细描述 HSEQ AP 以及案例公司不同供应商的 HSEQ 评估实践的现状。供应商公司的 HSEQ 管理采用了许多不同的方法，而案例公司选择的方法主要有：HSEQ AP 和工作后 HSEQ 评估，以及服务供应商的安全要求表格。这些供应公司在 HSEQ AP 方面的经验是值得鼓励的，但是 HSEQ AP 还需要进一步的发展。对安全指标和 HSEQ AP 的结果进行了研究，发现案例公司和供应公司的安全表现都有所改善，但同时也有许多其他保障安全的活动正在进行中。因此，尚不清楚 HSEQ AP 对这种积极发展到底有多大影响。此外，尤其是对供应公司而言，还需要对额外的安全指标进行研究，以更具体地证明实际的安全绩效水平。

第 6 章 港口场景的 HSEQ 框架介绍：综述和芬兰北部案例

摘要 本章论述了健康、安全、环境和质量（HSEQ）问题，并针对芬兰工业网络的 HSEQ 评估程序（AP）进行了详细的分析。详细讨论了在港口管理中使用 HSEQ AP 的必要性和实际应用的可能性。研究的港口位于巴伦支海北部。本章研究的重点是通过分析以前在造纸、化工和钢铁公司（过程工业）的供应网络中实施 AP 所获得的信息，以及使用实证访谈方法评估现有的标准和实践。本章研究的实证访谈部分主要在芬兰北部进行。在访问位于巴伦支海地区最大的俄罗斯摩尔曼斯克商业海港期间，还收集了一些关于 HSEQ 的信息。本章主要内容是基于文献和访谈工作的。前者包括对科学和工业文献的全面综述，后者是从海事公司（一个案例）的受访代表那里听取有关领域的 HSEQ 问题。这些方法的结合为我们提供了进一步的建议和方案。

第 7 章 驾驶室外卡车司机工作系统——宏观人机工程学发展方法

摘要 本地和短途（local and short haul, L/SH）司机除了在卡车驾驶室工作外，还在其他各种工作环境中工作。这些地点的环境和安全性差别很大。本研究将事故统计分析（过去）、视频观察（现在）和情景研讨（未来）三个时间角度分析方法结合起来，以提供可应用于交通运输业设计和管理过程开发的新方法。尽管新技术已经出现并融入以减轻驾驶员的工作，但在卡车驾驶室以外的环境中执行的工作仍然涉及体力活动、职业病危害和事故。因此，驾驶员在工作上的安全和工作能力问题仍然是一个需要持续、系统发展的领域。结果表明，为了成功地改善 L/SH 司机的工作环境，利益相关者的参与以及一套系统性方法是至关重要的。

第8章　危险液体运输中的致命非驾驶事故：芬兰工业系统中严重事故的调查程序及相关案例

摘要　本章研究的目的是论述芬兰对于致命事故的调查程序，并分析与危险液体运输有关的四起致命事故，尤其是卸货后服务操作相关的事故。运用内容分析法对调查报告中的事故因素进行识别。事故因素被分为两大类，对其中关于安全文化和安全态度的因素进行了讨论。还根据工作系统模型的五个要素，考虑了事故因素。最后，在芬兰的调查程序中，对事故的预防措施以简洁的形式提出。

第9章　芬兰两大用人企业员工失时事故总风险

摘要　本章针对芬兰两家大型工作场所雇员（$n=13000$）涉及的所有伤害性事故（10年内）总结了最新综合数据。本章研究旨在明确和评估不同失时工伤（LTI）事故（工作事故，家庭事故，休闲时间事故以及上下班途中的事故）中员工的风险。可以根据事故发生的频率和缺勤时间来揭示总风险。就所研究的案例（金属加工厂和市政机构）而言，家庭和闲暇时间事故似乎是数量最多的一类。对金属加工厂案例的初步分析表明，与白领员工相比，蓝领员工有更多的休闲性事故，同样他们在工作场所的事故也更多。

第10章 芬兰北部的HSEQ培训园区——建筑业创新合作论坛

摘要 建筑行业的安全培训需要新的方法和程序，以有效地减少目前事故的高发生率。HSEQ培训园区是一项新颖的安全培训创新方法，能够通过实际操作演练和主动参与来实现有效学习。该培训园区是近70家公司和社区合作在芬兰北部建造的。2014年春季，园区开始了培训活动。本研究概述了培训园区的设计、建造过程及架构。

第三部分 OSHM 效应

第 11 章 芬兰化学工业工作场所化学品的安全使用与风险防范

摘要 本章旨在探讨 HSE（健康、安全和环境）管理者和 OSH 员工代表如何看待以下议题：①化学品的安全使用，②预防优先级及管理措施，③在系统思维中 OSH 的集体氛围，④关于化学品安全与健康立法的论述。希望通过评估 OSH 的组织关系和技术措施来说明其潜在影响。49 名 HSE 管理者和 105 名 OSH 员工代表参与了在线问卷调查。在对 HSE 管理者和 OSH 员工代表的反馈分析中，预防优先级和管理措施对化学品的安全使用都有显著影响（$p<0.001$）。对于化学品的安全使用，我们发现以下影响因素：①培训；②职业安全与健康政策；③工作流程。在预防优先级方面，我们发现了以下影响因素：①技术；②评估和文件；③控制。在管理措施方面，有以下影响因素：①领导；②评估；③措施的利用。本章的定性结果发现，在遵守 OSH 法律要求、信息需求及不同政府机构之间的合作方面，情况有所改善。

第 12 章　芬兰化学工业的领导力和 OSH 过程

摘要　本章旨在探讨职业安全与健康中的组织和技术措施，并阐明立法、领导力、协作、预防、改进、监测、职业卫生保健（occupational health care，OHC）、培训和个人防护装备使用之间的潜在关系。参与者为化工公司的 OSH 管理人员（$N=85$）和 OSH 员工代表（$N=120$）。在回归分析中，领导力与协作在统计学上具有显著影响，而立法的质量同样在统计学意义上对持续改进具有显著的影响，此外，持续改进在统计学上有显著效果，员工培训在统计学上对预防优先性的确定有显著影响。另外，对工作环境的监测也具有显著影响，而领导力和协作对个人防护装备的使用具有显著的影响。在工作环境监测和风险预防的总体分析中发现以下四个因素：①技术和测量；②管理和指导；③监测；④风险管理。在动态过程中，例如在当下的化学工业中，采取命令和控制方法来实施指导方针进行自上而下的领导，这是不够的，还需要一个独特的具有自我调节能力的 OSH 系统（如责任护理规范）。

第 13 章　不同利益相关者对企业社会责任的探讨

摘要　本章旨在探讨芬兰劳动力市场不同参与者对企业社会责任的看法。芬兰工业联合会、芬兰工会中央组织和芬兰消费者协会各有一名高级专家被要求完成一项在线调查问卷。本研究的目的是阐明不同参与者如何看待企业社会责任。进一步研究的目的是收集对研究问题的各种反应，为今后的发展提供参考依据。主要的研究问题是：企业社会责任如何在商业活动中达到最佳实践？子问题是：企业社会责任是如何在以下日常活动中达到最佳实践的：①公司的商业战略；②金融家和股东；③客户和消费者；④公司的员工；⑤社区和政府。企业社会责任是企业与利益相关者相互承诺的战略过程。其目的是建立一种社会契约，最大限度地扩大所有相关者的福利。企业社会责任也是刺激企业积极行动改善公司健康、环境以及安全战略的重要途径。企业社会责任为公司提供了认识和有效应对社会挑战的能力。

第14章 芬兰企业可持续发展的案例研究

摘要 本章研究的目的在于探讨：①芬兰企业在可持续发展方面的各种实践；②如何将企业可持续发展应用于商业实践方面。进一步描述和解释企业可持续发展影响的跨部门关联机制、背景和结果模式。组织的价值观是决定优先事项的标准。有些企业价值观以道德为标准，例如指导决策的价值观，特别是指导流程的价值观。在企业可持续发展的框架内，价值观具有更广泛的含义。组织的价值观是管理者和员工做出优先决策的标准。企业可持续发展可以用不同的方法来实现：①在创造长期积极价值观和建立前瞻性程序时采用企业经营战略；②金融家和股东建立信托关系，实现长期可持续发展；③向顾客和消费者解释潜在影响和提供透明信息；④对公司员工要实施积极的措施，确保其作为一个完善的管理的一部分，形成良好的可持续性；⑤社区和政府通过提供相关信息，安排会议、公开听证会并承担咨询的做法，在产品、生产和服务的供应链中建立信任，并建立积极、透明的流程。

Preface

The "story" of this book started from the acronym HSEQ (health, safety, environment, quality). Some mail was exchanged between the editors and Springer about the key topics in HSEQ in the spring of 2012. An intensive one-day workshop was then arranged by the team of incoming editors to further develop the ideas in a more structured form. Quite soon, the editors' first key thoughts for a book were outlined in the form of a mind map (Fig. 1).

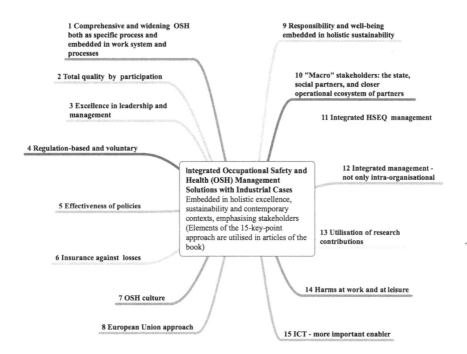

Fig. 1 General illustrated approach for the book

The writing process of the book chapters was planned and conducted, keeping in mind the mind map's set of contemporary trends and prerequisites, all of which are typical of the basic work approach of the editors as experts and actors in this field.

Figure 1 comprises the following points of emphasis, opened by a bit wider, though short way:

1. Occupational safety and health management (OSHM) is a specific sector of an organization's multidisciplinary expertise and management but simultaneously, it has to be "organically" involved in all managing activities of each workplace.
2. Each organization's entire resources contributing to holistic quality are needed, requiring broad and transparent participation at both individual and organizational levels.
3. Contemporary excellence in management strongly appreciates OSHM.
4. Appreciation of OSHM is based on both generic legislation and an organization-specific set of drivers for achieving competitive responsibility and public image.
5. The action of authorities and role of regulation try to widen OSHM's channels, to find synergies and, furthermore, evidence of the effectiveness of policies.
6. The insurance sector is an important institution touching and supporting all work organizations, as far as all risks, losses and asset management are concerned.
7. An organization strives to achieve its own OSHM culture, embedded as an essential element in its culture.
8. The European mainstream model has been strongly influenced by the history of Nordic countries' style of participation in tackling the challenges.
9. A cluster of driving forces, especially social responsibility, sustainability, well-being at work, productivity and general global contexts such as business, workforce and environment, encourages and even forces work organizations and key stakeholders to communicate and collaborate.
10. The key stakeholder role in all work organizations is played by the triple structure of the state, federations of employer associations and workers' unions. The management approach is supported and encouraged by the stakeholders' ecosystem, of which the mentioned triple entities are among the members.
11. Integrated risk management is an important concept; being linked with it is a key factor, concerning as well OSH issues and actors as a general good practice in management and finally, a desired goal of every citizen. A widely integrated HSEQ management system is an answer and a practical tool to handle future challenges.
12. The increasing importance of an organization's recognition of the network value and its influence in the development of commonly shared work systems—not only supply processes and "pure" business—is vital in the "new" OSHM.
13. Quantitative and qualitative, research-assisted development and innovation provide an increasingly significant opportunity and a stakeholder partner of work organizations.

14. Health, safety and well-being (at work, too) need to be cared for and maintained by each employee himself or herself, as well as by society on a 24/7 basis.
15. One key enabler for all of the above points is the expanding availability and utilization of information and communication technologies (ICTs), interfacing with people by means of desktop, portable and mobile devices. The digital operating environment is increasingly becoming a necessity in OSHM and HSEQ.

The chapter authors broadly represent universities, a research institute, the insurance sector and a public regulation body, and lastly, work organizations, companies and external stakeholders. Additionally, close industrial collaboration was guaranteed in many ways in the cases realized without any direct participation of companies. As the editors, we warmly thank all the authors presented in the list of contributors. Also, we express gratitude to numerous people who contributed to the book, organizations involved, their managers, experts, employees, entrepreneurs, partners, research and development (R&D) project organizations and key stakeholders. Finally, we appreciate Mrs. Maria Lindholm's skilful aid in processing the final phases of the manuscript.

We editors really hope that this book's "story" be continued by readers. The chapters are aimed to offer readers, both from industry and academia, new views and tools to go forward with OSHM.

Oulu, Finland, September 2014	Seppo Väyrynen
Vantaa, Finland	Kari Häkkinen
Helsinki, Finland	Toivo Niskanen

Contents

1   Introduction 1
     Seppo Väyrynen, Kari Häkkinen and Toivo Niskanen

Part I   Preventing OSH Losses by Developing Safety Management and Leadership

2   Safety Management—From Basic Understanding Towards
     Excellence 7
     Kari Häkkinen

3   Accident Sources and Prevention in the Construction
     Industry—Some Recent Developments in Finland 17
     Kari Häkkinen and Ville Niemelä

4   Application of Accident Information to Safety
     Promotion—Case Industrial Maintenance 25
     Salla Lind-Kohvakka

Part II   Views on OSHM and Cases Towards Continuous Improvements

5   Integrated Management Within a Finnish Industrial Network:
     Steel Mill Case of HSEQ Assessment Procedure 41
     Maarit Koivupalo, Heidi Junno and Seppo Väyrynen

6   Introducing a Scenario of a Seaport's HSEQ Framework:
     Review and a Case in Northern Finland 69
     Hanna Turunen, Seppo Väyrynen and Ulla Lehtinen

7 **Truck Drivers' Work Systems in Environments Other Than the Cab—A Macro Ergonomics Development Approach**......... 97
Arto Reiman, Seppo Väyrynen and Ari Putkonen

8 **Fatal Non-driving Accidents in Road Transport of Hazardous Liquids: Cases with Review on Finnish Procedure for Investigating Serious Accidents Within Work System**... 111
Henri Jounila and Arto Reiman

9 **The Total Risk of Lost-Time Accidents for Personnel of Two Large Employers in Finland**...................... 129
Liisa Yrjämä-Huikuri and Seppo Väyrynen

10 **HSEQ Training Park in Northern Finland—A Novel Innovation and Forum for Cooperation in the Construction Industry**.. 145
Arto Reiman, Olli Airaksinen, Seppo Väyrynen and Markku Aaltonen

Part III Effects of the OSHM

11 **Safe Use of Chemicals and Risk Prevention in the Finnish Chemical Industry's Work Places**...................... 157
Toivo Niskanen

12 **Leadership Relationships and Occupational Safety and Health Processes in the Finnish Chemical Industry**........ 185
Toivo Niskanen

13 **Discourses of the Different Stakeholders About Corporate Social Responsibility (CSR)**............................ 221
Toivo Niskanen

14 **A Case Study About the Discourses of the Finnish Corporations Concerned with Sustainable Development**........ 251
Toivo Niskanen

Part IV Conclusions

15 **Conclusions**.. 301
Seppo Väyrynen, Kari Häkkinen and Toivo Niskanen

Contributors

Markku Aaltonen Finnish Institute of Occupational Health, Helsinki, Finland

Olli Airaksinen HSEQ Park Northern Finland, Oulu, Finland

Kari Häkkinen If P&C Insurance Company Ltd., Espoo, Finland

Henri Jounila Faculty of Technology, Industrial Engineering and Management, University of Oulu, Oulu, Finland

Heidi Junno Outokumpu Stainless Oy, Tornio, Finland

Maarit Koivupalo Outokumpu Oyj, Tornio, Finland

Ulla Lehtinen Oulu Business School, University of Oulu, Oulu, Finland

Salla Lind-Kohvakka If P&C Insurance Company Ltd., Espoo, Finland

Ville Niemelä If P&C Insurance Company Ltd., Espoo, Finland

Toivo Niskanen Ministry of Social Affairs and Health, Occupational Safety and Health Department, Legal Unit, Helsinki, Finland

Ari Putkonen Turku University of Applied Sciences (TUAS), Turku, Finland

Arto Reiman Finnish Institute of Occupational Health, Oulu, Finland

Hanna Turunen Faculty of Technology, Industrial Engineering and Management, University of Oulu, Oulu, Finland

Seppo Väyrynen Faculty of Technology, Industrial Engineering and Management, University of Oulu, Oulu, Finland

Liisa Yrjämä-Huikuri Faculty of Technology, Industrial Engineering and Management, University of Oulu, Oulu, Finland

Chapter 1
Introduction

Seppo Väyrynen, Kari Häkkinen and Toivo Niskanen

If we think about an ideal contemporary approach to occupational safety and health management (OSHM), we can model it, for instance, with three cornerstones: legislation, standards and good practices. Of course, the third one comprises compulsory practices in one way or another, as well as a lot of non-compulsory ones. The drivers of the latter are typically OSHM-related measures for enhancing an organization's competitiveness. The OSHM-contributed competitiveness can result in internal or external outcomes. The internal ones include increased well-being at work and productivity, for example. The external ones can be recognized through an improved image noticed by clients, shareholders, stakeholders and citizens. Conformity with holistic quality is a crucial goal, both 'in and out'. This book predominantly covers the above-mentioned, wide variety of best practices.

As mentioned, the authors represent industrial companies, the insurance sector, universities, a research institute and a public regulation body. They provide diverse views with the common feature of realistic and direct contact with field practices, cases, many solutions, lessons learned and procedures for risk prevention. A wide variety of branches of industry is represented throughout the 13 articles.

The notion that organizations have core competencies in OSHM has been popular for much of the recent decades. The chapters of this book bring greater precision to the applications of OSHM (and not only the core competencies concept) by presenting a framework to help managers understand the OSHM system when they are confronted with necessary proactive measures to tackle the OSH

S. Väyrynen (✉)
Faculty of Technology, Industrial Engineering and Management,
University of Oulu, PO Box 4610, 90014 Oulu, Finland
e-mail: Seppo.Vayrynen@oulu.fi

K. Häkkinen
If P&C Insurance Company Ltd., Niittyportti 4, PO Box 1032,
00025 IF Espoo, Finland
e-mail: kari.hakkinen@if.fi

T. Niskanen
Ministry of Social Affairs and Health, Legal Unit, Occupational Safety and Health Department, PO Box 33, 00023 Government, Helsinki, Finland
e-mail: toivo.niskanen@stm.fi

challenges that lie ahead. When asking what sorts of proactive OSH measures their organizations are or are not likely to implement successfully, managers can learn a lot about capabilities in OSHM. Resources are the most visible of the factors that contribute to OSHM. The OSH resources can often be transferred across the boundaries of organizations much more readily than can OSH processes and values. Undoubtedly, access to high-quality OSH resources enhances an organization's capabilities in implementing effective, proactive OSH measures. The OSH resources are what managers most instinctively identify when the top management assesses whether their organizations can successfully implement the OSH measures.

The chapters of this book pose the challenge that business organizations do not have to wait for external pressures (e.g. legislation and standards) before they act to enhance their own HSEQ best practices. An important point to emphasize is that the best practices links among the organization, leadership, and individual managers and employees are not just about codes of conduct (e.g. in HSEQ management). Especially, best practices encompass a range of ways of communicating about HSEQ matters and promoting high corporate values in HSEQ. For example, these issues concern activities of HSEQ officers as part of the business production, OSH climate, training, monitoring systems, communication channels, socio-technical job design, information systems, etc. Selections of organizational strategies and policies in HSEQ have an impact on the values and practices that become accepted as 'the ways things are done around here'. Many of these elements come within what is known as organizational culture in HSEQ.

A wide variety of branches of industry is represented throughout the 13 articles, as follows:

Chapter 2 sums up the developments in the understanding of safety management in industrial companies, especially concentrating on features indicating good safety performance. Although the elements of good safety management are well known and reported in textbooks and standards, the real outcomes of safety performance are often far from excellent. The experiences show the need for holistic management as a basis for long-term development in safety improvement, rather than just implementing single tricks and solutions.

Chapter 3 reviews general health and safety hazards and the risks of accidents in construction work, including some accident statistics and a case description of a fatal accident. The results gained by the recent focus on improving safety performance by the Finnish construction industry itself and by its stakeholders are reviewed.

Chapter 4 summarizes some basic viewpoints on accident sources and risk assessment in industrial maintenance. A study focused on analysing severe and fatal accidents that had occurred during industrial maintenance in Finland. Additionally, a database of accident insurance institutions was revised to gain an understanding of less serious accidents in industrial maintenance. Some typical risk factors are poor system maintainability, defective task safety planning and shortcomings in customer cooperation.

Chapter 5 describes industrial service suppliers' HSEQ management practices in a general industrial context and in a shared workplace in the case company's premises. The case company is a steel mill in Northern Finland. The special HSEQ

Assessment Procedure (HSEQ AP) for suppliers is explained. The results of HSEQ AP, as well as other safety indicators, are analysed. The HSEQ AP cluster has achieved quite a significant size as of the autumn of 2014. The large, purchasing companies of the HSEQ AP cluster amount to seven; approximately 120 supplying companies have been assessed.

Chapter 6 addresses the need for and practical application possibilities of using the HSEQ AP in the management of a Finnish seaport. The port is situated in the northern Barents Region. The paper has document-based and interview-based sections. The empirical interview segment of this study is detailed and conducted mainly in Northern Finland. The information on HSEQ was also gathered during a visit to Russia's Murmansk Commercial Seaport.

Chapter 7 deals with local and short haul (L/SH) truck drivers' work in various environments (in addition to that of the truck cab), whose safety issues vary widely. This study combines methods from three time perspectives—accident statistics analyses (past), video observations (present) and scenario workshops (future)—to provide new knowledge that can be applied to the development of the design and management process in the transportation industry.

Chapter 8 reviews the Finnish investigation procedure for fatal accidents and analyses four actual instances related to the truck transport of hazardous liquids, especially in service operations after the unloading phase. Content analysis is applied to identify the accident factors from the investigation reports, for the purpose of finding prevention measures for future accidents.

Chapter 9 presents a new kind of comprehensive data on all injurious accidents (10-year time span) involving employees ($n = 13,000$) of two large workplaces in Finland. This study mainly aims to clarify and assess the significance of risks among employees in different lost-time injury (LTI) accident categories, namely accidents at work, at home and during leisure time, as well as when commuting to or from the work site.

Chapter 10 identifies an urgent need for a new kind of safety training in the construction industry. This training requires new methods and procedures if it is to effectively reduce the currently high incidence of accidents in Finland. A full-scale model of the industry's real work situations aims to utilize the HSEQ Training Park as a safety training innovation that enables learning via practical demonstrations and active participation.

Chapter 11 examines how HES (health, environment and safety) managers and the workers' OSH representatives view the following topics: (1) safe use of chemicals, (2) prevention prioritizations and measures of management, (3) collective climates of OSH in systems thinking and (4) discourses on legislation about chemical safety. Forty-nine HES managers and 105 workers' OSH representatives responded to an online survey questionnaire.

Chapter 12 explores organizational and technical measures in OSH, as well as clarifies the potential relationships among legislation, leadership, collaboration, prevention, improvements, monitoring, occupational healthcare, training and use of personal protective equipment. The respondents were OSH managers ($n = 85$) and workers' OSH representatives ($n = 120$) from chemical companies.

Chapter 13 examines how different stakeholders view various aspects of corporate social responsibility CSR. The main research question was: How does CSR become the best practice in business activities? The sub-questions were: How does CSR come to represent the best practices in the day-to-day activities of (1) the business strategies of companies, (2) financiers and shareholders, (3) customers and consumers, (4) employees of the companies and (5) communities and authorities?

Chapter 14 explores (1) the different kinds of practices implemented by Finnish corporations with respect to corporate sustainability and how (2) corporate sustainability is applied in their business practices. An organization's value is the standards by which managers and employees make prioritization decisions.

This book as a whole focuses on best practices, while considering legislation and standards. In this regard, we want to mention that the general, international standardized guidelines for OSHM under International Organization for Standardization (ISO)/CD 45001 will be published in 2016. Hopefully, the forthcoming standard will make OSHM more well-known and boost OSHM capability in work organizations. This book brings the reader beyond legislation and standards by giving relevant examples of work organizations that involve people and management. The question of how to manage organizations in practice by emphasizing social, economic and environmental sustainability and responsible management is answered in many ways.

Part I
Preventing OSH Losses by Developing Safety Management and Leadership

Chapter 2
Safety Management—From Basic Understanding Towards Excellence

Kari Häkkinen

Abstract The paper is summing up the developments in the understanding of safety management in industrial companies, especially concentrating on features that seem to be most prevalent for the companies demonstrating good safety performance. The review is based on the knowledge and experiences from the practice of industrial accident insurance as well as on the findings of the selected health and safety literature. The elements of success include e.g. good overall management, safety goals and competence requirements defined for managers, visibility and commitment by top management, rewarding and incentives for safe work, effective learning from accidents and incidents and continuous improvement using risk assessments, effective safety inspections and internal audits, as well as participative leadership practices including good communication and trust. The experiences show the need for holistic management as a basis towards the long-term development for safety excellence, rather than implementing single tricks and solutions. During the recent years, safety management systems in particular have been established in highly developed economies, thus improving the formal safety management approach. In the future, there seems to be a need to better tackle the more informal processes in safety, i.e. human behaviour, attitudes and safety culture. Furthermore, the complexity of health and safety call for the improvement of the general management theory to better tackle the complex human-centred issues in management.

1 Introduction

Despite continuous and ever-increasing volumes of organized action to prevent accidents and ill-health at work, injuries and occupational diseases still constitute a global problem in industries and services. Thus, we also get continuously new

K. Häkkinen (✉)
If P&C Insurance Company Ltd., Niittyportti 4, PO Box 1032,
00025 IF Espoo, Finland
e-mail: kari.hakkinen@if.fi

lessons from losses, often with major consequences both in human and in economic terms. The masses of information and knowledge of the losses and the risks to safety and health have dramatically increased during the recent decades, as well as the means to share the information and solutions for loss prevention. However, accidents continue to happen, although great progress has been achieved in many industries and occupations. What still needs to be done? This question is highly relevant to decision-makers, to business managers, to employees as well as to research and development institutions.

Safety efforts have been evolving from the early years of industry. The initial measures were directed to prevent major technical accidents in industrial processes by safety rules and disciplinary actions. Accordingly, new safety legislation was set in industrialized countries to force employers to organize safety and to take precautions and engineering against hazards. Accidents were considered to be caused either by unsafe acts of people or unsafe physical conditions of the nature or the technological systems (Heinrich 1959).

But further development of understanding revealed that each accident had rather multiple causes than a single cause. The evolution of systems thinking and multidisciplinary approaches such as ergonomics created new opportunities to deal with the complexity of the processes and events related to technological failures. Systems safety and risk management were developed to cope with the risks of major accidents in power and military systems as well as aerospace industry and high-risk industries more generally. The principles of safe man–machine systems are now to a large extent defined for all industries in technical standards and textbooks of machine safety, construction safety, chemical safety, etc. Risk assessment and safety technologies are standardized and applied today for everyday working tools and appliances.

Moreover, newer approaches in management science focused in human resources and change management, as well as the need for more holistic approach in management of the overall performance of enterprises, including health and safety. These developments together with the large scale technological disasters, e.g. Bhopal, Harrisburg, Chernobyl, Piper Alpha, Challenger, and more recently, the BP's US disaster cases have underlined the safety duties of top management. The growing concern of health and safety management has also led to further standardization of management practices in health and safety, e.g. OHSAS 18001 and ANSI Z10 (ANSI 2012; BSI 2007; Haight et al. 2014).

But in spite of all this development of formal legislation, standards and procedures, and increased management concern, there seems to be some missing links, and accidents at work have not disappeared. The remaining big issue seems to be related to the human nature. People still take risks at work, they are not following safety procedures, they do not communicate the safety messages, and foremen turn repeatedly a blind eye to unsafe actions. The motivational efforts of zero-accident programmes have been successful in many companies, but even prolonged zero-accident periods are ending to the next injury event. The 'Safety First' principle was created almost a 100 years ago during the early years of industrial safety movement. But in practice we still encounter repeatedly the 'Fail First' syndrome in health and safety development (Forck 2012). Accidents and injuries are bound into the culture of the work teams, enterprises and societies. Safety culture and understanding of the human element as a source of accidents need our consideration for the decades to come, both from practical and scientific point of view.

Management practices in health and safety vary greatly between companies. Industries have their own traditions and cultural approaches as to how safety issues are valued and decisions made to improve safety. Large international companies in high-risk industries have often strong safety cultures and strict management standards, with which all plants must comply. Within low-risk industries in terms of major accidents, often less attention is paid to safety issues. Basic health and safety legislation in each country covers all industries, whereas in practice companies take very different approaches to health and safety. However, the experiences of the insurance practitioners show that the successful companies in safety performance take often surprisingly similar actions and management approaches, regardless of the industry and type of business. The first decade of the 2000s shows a declining trend in accident frequency in large enterprises in many industries. The accident rate differences are shrinking between industries and some industries with high accident rates have achieved major improvements, in particular some metal and construction industry companies (see Fig. 1). The improvement of safety is gained by the increase in management concern to safety, including the focus on safety management systems and new occupational safety legislation requiring more attention to health and safety risk management. This trend of safety improvement is not visible in the overall national work injury statistics, thus it seems that small- and medium-sized companies have been more stable in their accident outcomes.

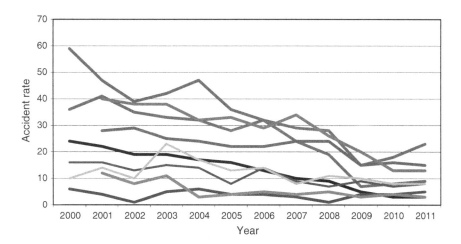

Fig. 1 Accident rate development in nine major industrial companies representing different industries (construction, metal, power, chemical) in Finland during the time period 2000–2011 (*Source* Statistics of If P&C Insurance Ltd for injuries more than 3 days absence from work, accordingly with the national and EU/Eurostat criteria)

The purpose of this paper is to sum up the most important management issues which repeatedly seem to have most influence on success in safety performance at a company level.

2 Health and Safety Goals and Targets Clearly Defined and Managed

'What gets measured, gets managed'. This well-known statement of management guru Peter Drucker is highly relevant in the case of health and safety. Unlike other major issues in company performance, such as production volumes and profit, health and safety is to a last extent an intangible issue. Therefore, tangible and concrete goals are important in spite of the fact that they do not exactly reflect the overall health and safety sphere of required action. An essential management task lies in defining clear goals and targets, from which key performance indicators applicable to follow-up are then derived. Without goals and targets, well-intentioned statements written in safety policies are just dead letters.

Accident frequency rates are often the first measures of interest. They can be combined with severity rates to yield a relevant risk index. Cost indicators are also interesting to companies—the costs may be expressed by the insurance premium rates or directly by the real accident and occupational disease costs for the company. The absence rates due to accidents and sickness reflect in many ways the health and safety performance, being increasingly popular in industry. Companies with good safety performance have usually safety metrics with many indicators, they may record, e.g. numbers of conducted safety talks and audits, training days, and near-accident report numbers and actions succeeding them. Thus, a good pattern of safety targets includes both leading and lagging indicators to be followed. It is also a good management practice to integrate health and safety goals into the business goals and strategies, and the safety targets are expressed in the key performance indicators of the company.

3 Inclusion of Health and Safety in Management Competencies

Expectations set on competence are usually managed by the human resource function and written into work descriptions and listing of the personal duties. If health and safety is not included in the competence specifications of managers, they may fall outside the focus areas of the mindset during the daily work. Combined with a lack of fixed goals and targets, this will likely lead to a situation in which managers pay only marginal attention, if any, to safety issues. Relying on legal pressures from health and safety legislation is insufficient in confirming management's commitment.

In addition, health and safety issues should be essential elements in performance appraisals. In general, safety excellence seems to be related to good overall leadership, thus giving an additional measure of success in management as a whole. Many companies include health and safety results in salary bonus payments at management level. When management duties have been planned and integrated in job specifications, managers have a strong motivation to the safe action of their subordinates and fellow employees.

4 Management Commitment and Visibility in Health and Safety Actions

People are often very astute in discerning management's real commitments. They soon may take note of whether or not management is genuinely interested in the health and safety of employees. When top management participates in company safety meetings and discusses safety goals in connection with production targets, employees receive a clear signal that safety is an important and valued issue for management. 'Safety first' is not just a slogan, its actual meaning is checked in all decisions, meetings and encounters with management.

In very good companies, CEOs increasingly participate in safety walks and talks, or conduct such actions independently. Managers also need to follow the safety rules, using hard hats and other personal protective equipment as required by the company policy, following the speed limits specified when driving at the premises of the company site, etc. It is important to walk the talk, to be visible, to be present and engaged.

5 Rewarding Safe and Injury-Free Work

'What Gets Measured and Rewarded Gets Done'—this is the way in which renowned safety guru Dan Petersen (1996) has modified the Drucker statement. He thereby implies that rewarding is an important element in effective goal-setting. Safety is largely a matter of intangibles, and the results are traditionally noted, if at all noted, mainly in retrospect. The notions tend to be related to failures, i.e. accidents and costs, rather than success. It is therefore important to render safety tangible and create incentives for proactive, forward-looking and positive measures.

Rewarding achievements such as an injury-free year, the most active team in near-accident reporting and the best innovation solutions to improve safety, can generate a great deal of positive energy and actions in organizations, thereby further reducing accidental losses. Both short-term and long-term achievements need consideration. In high-risk occupations, having an injury-free month can be worth awarding, while it is also important to notify the continuous improvement over the years.

6 Immediate Management Action Against Risk-Taking and Unsafe Acts

When an employee is violating safety rules or working without the required personal protection, foremen and co-workers alike should act immediately to correct such poor practices. In companies with a good safety culture, actions against violations, whether intended or unintended, are regarded as a positive challenge to learn and improve. But bringing the culture up to this level is not easy. Foremen and management need coaching. They should have the ability to communicate with and motivate employees individually, while understanding the interlocutor's point of view. It is often necessary to overcome one's own complacency first. Finally, it remains the duty of management to take disciplinary action to correct a situation, if necessary. On the spur of the moment, it may feel easier to turn a blind eye. In a strong safety culture, leaders are encouraged and trained to act immediately. Inaction is more dangerous in the long run, since it reinforces a risk-taking culture, making future improvements more difficult.

7 Regular and Active Communication and Discussion of Health and Safety

Constant and active dialogue between employees and management is a precondition for continuous improvement in health and safety. When the minimum legal requirements are complied with, and the company is striving for further progress to safety excellence, issues are less and less often resolved through absolute yes-or-no answers. There will be opinions for and against. Sometimes improvements here will cause more difficulties there. In good companies, we find ongoing conversations in the search for better solutions, in order to improve safety and well-being at work. Good leaders are also good at listening and communicating. In the best companies, the accident outcome may be close to zero injuries, while thousands of safety talks and conversations are conducted by management.

A number of experiences have shown that in Nordic enterprises, there are fewer accidents in Swedish plants than in Finnish plants engaged in very similar production activities. There may be several underlying reasons for this, but the Swedish culture and tradition of thorough discussion and communication may be a factor. The management style in Finland has promoted straight decision and action rather than dialogues with employees.

Spangenberg et al. (2003) compared accident rates between Danish and Swedish working teams in the construction site of the Öresund Link bridge. They found that the Danish LTI accident rates were about fourfold the rates of the Swedish workers. They concluded that the difference was mainly due to individual and team level differences, i.e. education, experience and attitude. While there is some evidence that the differences are partially rooted in the basic education and cultural valuation of the construction professionals, it is also an indication that the management culture, in particular the communication and feedback at the foreman level, are strong explanatory factors in safety performance. The results of Hyttinen (1994) also support the meaningful role of the supervisor communication for successful safety performance.

8 Safety Inspections and Risk Assessments

Internal safety inspections are important features in good safety practice. In daily production tasks, it is not always possible to detect all safety deficiencies. It is also well known that experienced employees can be too familiar with their work to note even imminent dangers. Regular walk-through surveys are therefore required. Safety checks may be conducted weekly, monthly or with some other level of frequency, depending on local risk conditions. It is usually advisable to use a checklist form appropriate to the plant, in order to conduct a systematic consideration of all of the relevant hazards. In the case of advanced safety inspection routines documented and issued using company systems, there is also a communication tool and checkout for the implementation of corrective measures by management.

Risk assessment procedures in health and safety are legal requirements in most countries and also essential elements of the safety management systems, being they certified or not. In formal risk assessments, the risks are listed based on their priority in terms of severity and likelihood of the possible accident. In spite of the fact that there is a subjective element involved in such estimation, the process will yield a rational basis for decisions to act on hazards encountered. Further, the follow-up assessments are giving estimates of residual risks for those risks counteracted, to complete the prevention with additional measures where needed.

Safety is at first hand a responsibility of business and production management, rather than the duty for safety managers or specific safety organization. In good companies, we increasingly see that risk assessment and safety walk-through inspections belong to line management duties. Safety function has an important advisory role for management, and to secure that the risk management process is functioning the way it was meant to.

9 Investigation and Reporting of Accidents and Near-Accidents

An accident at work is always an opportunity for learning, for management as well as employees. A good accident investigation routine records what happened, why, and lessons on how to prevent the accident's recurrence. For an employer, accident investigation is also a legal duty, as well as being necessary in order to begin an insurance claims process in workers' compensation systems. It is usually not enough to check immediate causes such as carelessness or poor housekeeping. Understanding the root causes of an accident often reveals new paths for prevention, e.g. in process improvement, management actions and safety culture improvement.

A range of tools are available for root cause investigations, ranging from simple questioning techniques such as '5 x Why', to sophisticated software packages. In well-run companies, information on accidents is distributed openly throughout the organization by 'hazard-alert' and similar illustrated information reminders, in order to remind employees of risks and precautionary measures in the workplace.

Activity levels in near-accident reporting and accident outcomes seem interconnected. We have found that, in companies that have achieved major increases in their reported near-accidents and hazard observations, a clear decline in workplace injuries has resulted at a same time. Knowledge and experience are dramatically bolstered by advances in the reporting of minor incidents. Active reporting of near-accidents is also an indicator of a good safety culture, where people dare to talk and report their own failures without fear, and a near-accident note is seen as a positive reminder for safeguarding colleagues at sites and as a sign of management caring of the employees.

10 Good Overall Management

Good overall management, including good communication both top–down and down–top, seems to be a precondition for success in safety and health. It also corresponds to the management approach required for good business results. Enterprises that respond positively based on all of the above-mentioned criteria are likely to be among the best health and safety performers in their industry. Moreover, the lessons from major accidents also highlight the needs for a good overall management. Trevor Kletz (1994) concluded his discussion on the 1988 Piper Alpha disaster, that no single act or omission was responsible for the accident. And he summarizes: 'Perhaps the most important lesson of all that can be drawn is that the sum and quality of our individual contributions to the management of safety determines whether the colleagues we work with live or die'.

Health and safety management systems have improved the overall understanding and implementation of good management practices. However, a functioning formal management system is not enough. There is a need for a more personal approach to health and safety. Each and every individual must look in the mirror and ponder their personal attitudes and values. Safety is much related to the emotions, attitudes and values of people. Besides following the rules and regulations, it is about caring and feeling. More thoughtful and lively dialogues are needed on how to improve our mindsets and safety practices, in order to get all fellow employees, teams and managers engaged in coordinated actions to move the safety culture towards the achievement of an injury-free workplace (see e.g. Choudry et al. 2007; Hudson 2007).

This is also a challenge from scientific point of view. More understanding is needed regarding safety culture, human behaviour and safety leadership in the contexts of businesses and industries, to better manage the risks in health and safety in future. Moreover, it may be that the general management theory needs some further development in order to better cope with the complexity of the issues involving human nature and behaviour in organizations and related to safety (see Carrillo 2011). This complex nature of human relations is a challenge for management not only in safety, but also more generally in business management.

References

ANSI/AIHA/ASSE (2012) Occupational health and safety management systems (ANSI Z10-2012), Des Plaines

British Standards Institute (2007) Occupational health and safety management systems: requirements (OHSAS 18001:2007)

Carrillo R (2011) Complexity and safety. J Saf Res 42:293–300

Choudhry R et al (2007) The nature of safety culture: a survey of the state-of-the-art. Saf Sci 45:993–1012

Forck MA (2012) What safety leaders do. SafeStrat LLC

Haight J et al (2014) Safety management systems—comparing content and impact. Prof Saf 5:44–51

Heinrich HW (1959) Industrial accident prevention, 2nd edn. McGraw-Hill, New York

Hudson P (2007) Implementing a safety culture in a major multi-national. Saf Sci 45:697–722

Hyttinen M (1994) The successful leadership behavior of the building construction site supervisor. Dissertation, The University of Oulu. Acta Univ.Oul.C 78, Oulu (in Finnish)

Kletz T (1994) Learning from accidents, 2nd edn. Butterworths-Heinemann Ltd, Oxford

Petersen D (1996) Safety by objectives—what gets measured and rewarded gets done, 2nd edn. Van Nostrand Reinhold, New York

Spangenberg S et al (2003) Factors contributing to the differences in work related injury rates between Danish and Swedish construction workers. Saf Sci 41:517–530

Chapter 3
Accident Sources and Prevention in the Construction Industry—Some Recent Developments in Finland

Kari Häkkinen and Ville Niemelä

Abstract The construction industry is well known as one of the riskiest industries in regard to work accidents, even in advanced economies. This paper reviews health and safety hazards and the risks of accidents in construction work. Some accident statistics are presented based on the claims data of the Finnish workers' compensation system, with a narrative case example of a fatal accident. The fatal accident demonstrates the risks involved with working at height and the hazards, which emerge at a shared workplace where several sub-contractors work together. The recent focus on improving safety performance has contributed to positive developments, which have resulted both in an improvement in the claims statistics as well as a number of new innovations to improve safety practices at construction sites. Some large international contractors have taken the leading role in the development of safety management in construction projects. The improvements seem to be strongly related to the sharpened role of top management, as a consequence of which there is now increased attention paid to safety by supervisors and employees alike. New tools and processes, i.e. in incident investigation and reporting as well as work planning procedures, have contributed to more effective and systematic site safety practices. However, the way forward towards a zero-accident culture in the construction industry is a continual and persistent effort through which a significant improvement in the results may occur. This means that, awareness by all of the parties involved is a constantly needed prerequisite, and complacency should be avoided.

K. Häkkinen (✉)
If P&C Insurance Company Ltd., Niittyportti 4, PO Box 1032, 00025 IF Espoo, Finland
e-mail: kari.hakkinen@if.fi

V. Niemelä
If P&C Insurance Company Ltd., Niittyportti 4, PO Box 1022, 00025 IF Espoo, Finland
e-mail: ville.niemela@if.fi

1 Characteristics of Health and Safety Hazards and Risks in the Construction Industry

Construction projects differ vastly in the scale and the nature of work involved, i.e. they can range from the building of houses or commercial and industrial buildings and plants, or heavy civil engineering constructions such as bridges, tunnels, roads, water systems to railroads, harbours and airports. Furthermore, the hazards connected with repairing older buildings and structures are different compared to the construction of new houses and infrastructure. The demolition of old buildings is generally considered to be more risky for employees than working on new constructions.

There are very often many employers working together at the same construction site. A large construction project typically has a main contractor and a large chain of sub-contractors, including many small employers with a variety of competencies and skills. The safety management of such large construction projects is very challenging. Good cooperation and communications between all parties concerned is of utmost importance. While the main contractor has the major role in the overall safety management and coordination, it is important that all employers and employees alike actively take care of their duties in regard to matters of safety. Moreover, clear definitions of the roles of the clients and the owners of the constructions are nowadays prescribed to be essential for loss prevention in major construction projects.

Typical accident scenarios on construction sites are slips and falls whilst walking and working on sites, falls from a height, motor vehicle collisions, excavation accidents, injuries involving the use of hand tools and machines, and being struck by falling objects. Some of the main health hazards on-site are solvents, noise, asbestos and manual handling activities. In specific tasks and projects, construction employees may be exposed to special hazards which could lead to serious accidents such as electrocution, drowning, asphyxiation, fire and explosion (ILO 1998).

Howarth and Watson (2009) have classified the hazards of construction projects in the following five categories: (1) Hazards presented by the local environment; (2) Hazards presented by work activities; (3) Hazards presented by a deficiency in people's knowledge, attitude and behaviour; (4) Hazards presented by the movement of people, plant and machinery; (5) Hazards presented by materials. The identification of safety hazards is a key process of safety management exercised throughout all phases of a construction project. Due to the complex nature of the construction safety issue, specific laws and regulations have been issued for construction sites in most highly developed countries. These laws and regulations supplement the general employer duties described in the general health and safety legislations of each country.

2 Some Loss Statistics

In the construction industry, the frequency of accidents is higher than in most other industries, and a major share of serious and fatal accidents is every year registered to construction work.

However, during recent years, a positive declining trend of accident frequency rates is visible in the statistics for the construction sector in Finland, thus indicating a positive safety development (see Fig. 1).

According to the insurance statistics of Finland for 5 years (2005–2009) the most common working task performed before an injuring event occurs in the construction industry is walking and moving at a worksite (Fig. 2a). The share of human movement injuries, in serious accidents leading to permanent work disabilities, equates to more than half of the cases. Injuries to shoulders represent as much as 25 % of the accident pension cases and the total amount of injuries to the upper limbs equates 41 % of accident pensions (Fig. 2b). In addition, accidents occur frequently when using hand tools, in the manual handling of materials and during the use of machinery. The most common body parts injured are hands and arms—from fingers to shoulders, and legs including knees. The fatal injuries result mostly from head injuries or multiple injuries to the body due to falls from a height.

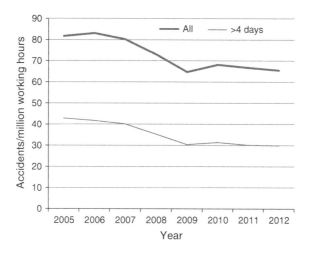

Fig. 1 The trends in accident frequency rates in the construction industry of Finland during the years 2005–2012, presented for all compensated accidents and for of the accidents leading to more than 4 days' absence from work (The data are issued by the Federation of Accident Insurance Institutions in Finland)

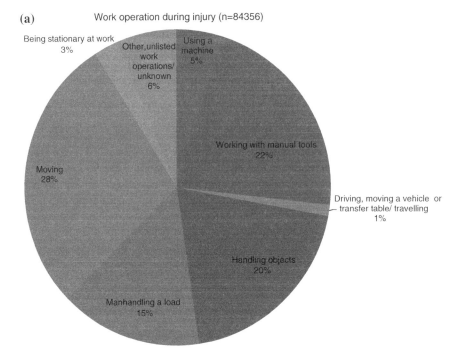

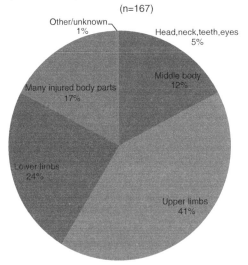

Fig. 2 a All notified accidents in the construction industry by work operation of the injured person leading to the injury. **b** Injured part of the body in serious accidents leading to pension in the construction industry (*Source* The national statistical database issued by the Finnish Federation of Accident Insurance Institutions, including all notified work accidents in Finland during the years 2005–2009)

3 A Fatal Accident Case: Working at Height

Working at height is work where a person could fall and get injured if adequate precautions are not taken. This description applies to most industrial workplaces and is—and has been—a major source of many fatal accidents, particularly in the construction industry. The case below is a classic situation that led to a fatal injury when falling precautions failed.

3.1 What Happened?

A three-floor enlargement project of a school building was ongoing and several different tasks were under construction simultaneously. Two men were installing a temporary plasterboard wall on the third floor when they ran out of boards and were instructed to carry the missing pieces from the second floor after their lunch break. During this break a group of sub-contractor workers started installing permanent handrails in the stairway where the boards where supposed to be carried through.

As a part of this task the temporary handrails from the landing on the third floor were removed and were still missing when the workers carrying the plasterboards were passing the spot. Whilst turning the board to enable it to go through a doorway on the right, the worker carrying the board from the rear apparently stepped off the landing and fell 3.5 m. He died after 3 months in hospital care.

3.2 What Caused the Accident?

The installation of the handrails created an immediate risk of falling for anyone in the area. The coordination of tasks during production planning should have taken these hazards into consideration and scheduled them so that the temporarily dangerous access would not have been needed until it was safe again.

A sub-contractor creating a safety hazard is responsible for arranging and completing the required safety measures. Because the temporary handrail needed to be removed in order to complete the task, the obvious danger of falling should have been notified clearly and the access to the stairway and landing closed until the handrails were completed.

Carrying the plasterboard through a stairway was not the normal procedure, the other boards were lifted earlier to the third floor with a telescopic handler, but this option was not available anymore. In any case, the stairway was used as the main route to the third floor. Closing the access during the handrail installation work would have been the absolute correct thing to do.

3.3 Preventing Similar Accidents

The main contractor is responsible for coordinating the construction project activities so that separate but simultaneous tasks do not create hazardous conditions for anyone on the same construction site. The main contractor must also appoint a safety coordinator to overlook and verify the correct safety actions, as well as the ones that the sub-contractors are responsible for.

In this type of situation where a work operation creates a high-risk condition in a certain area, the area must be secured so that no one can enter without proper safety precautions. It is also important to assure the safety of workers working in a high-risk area. Adequate falling protection, a safety harness in most cases, is vital when work must be done in a place without stationary fencing.

In shared workplaces like the construction site in this case, appliance of the same safety rules and procedures regardless of contractor would benefit all. In this work situation during lunch break, the installation workers stepped aside to give room to pass when workers carrying the boards entered the landing. So despite the hazard, the mentality was more "everyone takes care of his/her own work" rather than "safe working together". When all workers on a site know and follow the rules and procedures, the evolvement of a safety hazard into a serious accident can be stopped before it is too late.

4 The Focused Prevention Approach Towards Zero Accident Performance

While work accidents regularly have multiple causes that can be identified by a thorough accident investigation process, some general sources of loss can be identified based on the long-term experience with construction safety issues:

- *Inadequate access systems, walkways and fall protection at worksites*, especially in the use of ladders, scaffolds and other temporary solutions, unprotected openings on floors and the lack of guardrails, personal fall protection equipment not in use.
- *Poor housekeeping and tidiness of worksites* contribute to numerous accidents at construction sites, e.g. the storage of materials and tools, the installation of hoses and cables, dusty, oily, dirty and icy floors, inadequate space for walking and working due to litter and extra materials at the sites.
- *Safety management systems are not working*, this may be due to inadequate organization or poor leadership and supervision at a site, or because the agreements and communications between the contractors are not observed.
- *Planning of the construction project* is inadequate, with a lack of proper risk assessment procedures and failures in the control of deviations from the original planning.

In recent years, the construction industry in Finland has actively developed new innovations to improve safety performance. The new approaches in prevention have been achieved to a large extent with close cooperation between employers, trade unions, occupational health and safety inspectorate and insurance companies. At present, an ambitious goal to reach a zero accident level for the industry by the year 2020 has been published. Large international construction companies have a leading role as the forerunners in this development. Moreover, the interest of top management towards safety improvements has grown a lot during recent years. Nowadays, accident frequency rates and trends constitute a major performance indicator for construction companies, and employers compete positively with each other for safety excellence. Parallel to this development, line managers, foremen and employees are currently the main actors in health and safety, rather than only safety managers and specialists.

The sharpened focus on safety management at construction sites has also improved tools and processes in practical health and safety work. The traditional weekly inspections of construction sites have been substituted by a more systematic TR observation method (European Agency for Safety and Health at Work 2004). The TR tool gives a measurement indicator based on the observations of housekeeping and other safety-critical issues at the site. Thus the progress of the site in health and safety performance can be followed week by week by the company management and safety inspectors. More recently, iPads and mobile phone tools are used to aid the assessment, thus making the process more effective and repeatable at site conditions.

Accident investigation routines have been improved particularly by the large contractors. The trends are towards more detailed fact-finding investigation as well as interviews to reveal the root causes of the incidents. The pressures from management and clients of the projects as to better safety concern have also improved the reporting of the near-accidents at construction sites. The best construction companies now report some thousands of near-accidents and at-risk situations, while only a few lost-time accidents have occurred at their sites.

It is a well-known fact that accidents at construction sites often occur at work situations which are unplanned, often due to unexpected changes in the work flow. To improve the site practices, a major Nordic contractor Skanska initiated the so-called Work Safety Plan. In this procedure, employees and management together plan a reasonably safe solution to an encountered new situation without any pre-existing plan. The plan is documented and signed by both an employer and employee representative. The procedure has been used very frequently, and thousands of plans are documented yearly at the Finnish work sites. But the challenges are still ahead: in too many accident cases still occurring, no need for Work Safety Plan has been identified before the start of the work.

The way towards a zero-accident culture is long and difficult in the construction industry. Parallel to declining accident rates, the improvement gets more challenging, and from time to time there may appear increasing of losses. Prevention is first and foremost a human affair. Both managers and employees need to be continuously awakened and aware of the hazards and develop new ways to improve safety, to communicate better, and to take care of the daily routines as to the maintenance of the good working conditions. Construction sites are changing continuously and people are exposed to the changing conditions, be it the weather, technology, organization or the economic environment, or something else. But the building companies and their employees are now more and more committed to the goals to improve safety. The way towards zero-accident culture in the construction industry is a persistent effort where results may be meaningfully improving, but awareness of all parties involved is needed constantly, and complacency should be avoided.

References

European Agency for Safety and Health at Work (2004) Achieving better safety and health in construction (information report). Office for Official Publication of the European Communities, Luxembourg

Howarth T, Watson P (2009) Construction safety management. Wiley-Blackwell, Chichester

ILO (1998) Encyclopaedia of occupational health and safety, vol 3, 4th edn. Industries and Occupations, Geneva

Chapter 4
Application of Accident Information to Safety Promotion—Case Industrial Maintenance

Salla Lind-Kohvakka

Abstract Maintenance operations are conducted basically in all industrial environments. The risks vary greatly, depending on the operating environment and conditions, and on the task involved. This article summarizes some studies that have focused on industrial maintenance. The studies were conducted during the years 2004–2014. The main materials have been gathered from various accident databases, and they are supported with risk assessments carried out in companies. A major study has focused on re-analysing fatal and severe non-fatal accidents that have occurred during industrial maintenance work in Finland. In addition, the database of the Finnish accident insurance institutions was revised in order to gain an understanding about less serious accidents, which occur in industrial maintenance. Some typical risk factors are poor system maintainability, defective task safety planning and shortcomings in customer cooperation. Risk assessment and management should pay attention to varying tasks and operating environments, as well as to careful task safety planning. This article summarizes some basic viewpoints about accident sources and risk assessment in industrial maintenance.

1 Industrial Maintenance and Risks

Maintenance and service are a group of activities that are relevant for all industries. In 2012, the sector employed 22 000 persons full-time in Finland (Statistics Finland 2014). In addition to this group, the number of workers conducting maintenance operations part-time or occasionally is probably much higher. In this document, the terms "maintenance" and "service" refer to the following activities:

- preventive maintenance (i.e. scheduled disturbance- and failure-preventive operations) and after-sales service

---

S. Lind-Kohvakka (✉)
If P&C Insurance Company Ltd., Niittyportti 4, PO Box 1032, 00025 IF Espoo, Finland
e-mail: salla.lind@if.fi

- unscheduled operations, including troubleshooting and corrective maintenance
- calibration and testing.

In addition, the recent statistics in this article include also the installation of machinery and technical systems, as they are included in the group of maintenance activities in Finnish statistics.

Maintenance operations in industry include a variety of risks, which may endanger the maintenance worker during the maintenance task and/or weaken the post-maintenance system's safety and reliability (Lind 2009). For the worker, maintenance operations may be riskier than many other tasks in industry, because in addition to the actual maintenance operation itself, maintenance operations also include system disassembly and reassembly with the related risks.

From the viewpoint of accident prevention, industrial maintenance has inherently certain challenging features. Depending on the industry, there are various hazards relating to operating environments. In addition, the operations with the related risks can vary accordingly. In some cases, both the operating industry and the operations may be new to the worker. Risk management is often approached on the basis of the industry, i.e. the assessment methods and statistical examination are chosen and/or focused on the basis of the industry. In the case of maintenance, the approach to risk management is activity-based and industry-independent. As maintenance is an exception from the normal use of a system, the regular risk assessments conducted in companies may not be extensive enough to point out the possible hazards and hazardous events. At the same time, the system and machine maintainability may have been designed poorly and/or the system or some parts of it may have changed over time so that some old risks have been eliminated while some new ones have been created.

While many other industrial operations are moving from the factory floors to control rooms, maintenance operations are (and probably will be also in the future) carried out on the human–machine interface. Meanwhile, the tasks are becoming increasingly complex, as the industrial systems are more and more automated but still rarely designed for maintenance. In Finland, maintenance is often outsourced, which can also contribute to safety losses, as the site management may be distant and the responsibilities unclear. Also cooperation with the customer (and other operators) on site requires careful planning and management because the operations are carried by the customer and maintenance workers in the same premises but with different aims and responsibilities.

Figure 1 (quoted from Lind 2009) describes a schema of maintenance operations, including the development of the need for maintenance (on the left) with possible unwanted consequences. These can be environmental emissions and/or impaired safety. The right side of the schema summarizes the main stages of a maintenance operation. Other main components in the schema are the direction of the analyses: in the event of an occurred accident the analysis has to be top-down (e.g. fault tree analysis). In order to identify possible accident contributing factors beforehand, the analysis can be started from the unwanted outcome and then the model chain of contributing events leading to the accident (Lind 2009).

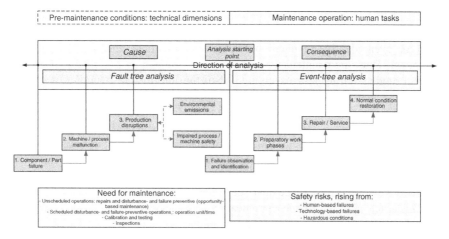

Fig. 1 Maintenance: cause–consequence schema

The unwanted outcomes in maintenance operations can be grouped into two main categories: (1) technical malfunctions, which may weaken system safety and reliability, and (2) occupational risks harming the worker's health immediately or over time (accident risks, ergonomics risks, exposures). The technical malfunctions can be the result of human error during a maintenance operation. For example, a part can be replaced with the wrong component, the component can be installed incorrectly into the system, or the system can be reassembled in the wrong way. This kind of human error has been explored in various publications, such as Kumamoto and Henley (1996), Reason (1997), Dekker (2006), and Reason and Hobbs (2003). The technical malfunctions can result also from component and system failures or breakdowns.

Occupational risks can be immediate accident risks occurring during the disassembly, the actual maintenance task and/or the system's reassembly. Such risks can be caused by, for example, physical or chemical factors, or cognitive load. According to Reason (1997), accidents have their pathways in three main groups, i.e. human actions, local workplace factors and organizational factors. The two latter ones are latent conditions, whereas unsafe acts are often referred to as active errors. In the case of maintenance, this grouping can be divided so that the safety risks are grouped into human-based failures, technology-based failures and hazardous conditions. These can lead to an immediate accident/incident during the maintenance operation. The first two can also become latent pathogens in the technical system, so that they cause problems sooner or later after the maintenance operation has been completed. In addition to the risks endangering health and safety directly, there can also be accident contributing factors that do not cause accidents as such, but can accelerate and accumulate the risks of accidents. Such accident contributing factors can rise from human actions, technology-based failures (or features leading to poor maintainability), and/or hazardous conditions (see e.g. Lind 2009).

Maintenance operations are highly dynamic and complex, especially when it comes to fault identification and troubleshooting. The tasks may be impossible to plan or even identify in detail beforehand, if the fault and/or its location is unknown. Thus, the maintenance worker(s) must be prepared to change plans and mental models during task execution. Also, the information regarding necessary tools, risks and safety measures must be updated during task execution. Qureshi et al. (2007) proposed a systemic approach to the identification and management of accident causes, while designing systems to be resilient and error tolerant. The systemic approach supports also the idea of managing risks in a dynamic complex system, which fits well the idea of managing maintenance-related risks. However, the problem may eventually be the difficulty to identify all possible actions and operations with the related accident scenarios.

This article aims to provide a practical view to accident risks and risk assessment in industrial maintenance. The focus is on the accident sources, based on real accident data. In addition, the article provides information regarding risk assessment and management in practice.

2 Methods and Materials

This article is a review, going back to two separate studies and research work in Finnish workplaces. The studies were conducted during 2004–2014. The focus of the studies and research work has been on maintenance operations and maintenance workers. The first study, with the focus on risk management in industrial maintenance, was reported in Lind (2009). The study made an extensive review of accidents, risks and risk management in Finnish companies providing maintenance and after-sales service. Some key findings regarding e.g. accident sources in the case of fatal and severe non-fatal accidents were presented in Lind (2008). The main data for the study included all workplace fatalities in industrial maintenance in Finland, which had been reported between the years 1985–2004. In addition, the study made a review of the types and sources of all severe non-fatal accidents, which had been reported during the years 1994–2004. The access to these two accident report databases is public. Other methods in the study included risk assessments in maintenance providing companies, which have a total of 15 customer sites between them.

The findings are completed with the second study, which focuses on statistics and real accident data from the viewpoint of an insurance company. A major database is provided by the Federation of Accident Insurance Institutions in Finland (FAII). The database provides numerical background information from real accidents, supplemented with accident descriptions. This database includes all of the accidents in Finland which have been reported to the Finnish accident insurance institutions. Access to the database is primarily granted to insurance institutions only. Statistics Finland hosts the open database of general statistics describing e.g. key numbers in Finnish industry. In this article, the referred accident cases are summarized from the (FAII 2014).

3 Accident Statistics and Typical Accident Scenarios in Industrial Maintenance

According to the statistics, the total number of workplace accidents in maintenance operations has been decreasing since 2007. The number of persons employed full-time in industrial maintenance is currently about 22 000. It has increased slightly during the past decade, which may be the result of increased outsourcing. During the past few years, the number of accidents has varied from between 605 (year 2009) and 998 (2007) (Statistics Finland 2014).

The statistics indicate that the minor accidents leading to 0–3 days of absence are typically caused by materials, splinters and products (FAII 2014). A typical accident of this type is a minor eye accident, caused by a splinter or fragment released during material handling (e.g. welding). Other major injury causes involve hand tools, which may cause e.g. minor cuts and bruises. In general, accidents in maintenance operations can affect any body part, and the accidents may occur at any stage of disassembly and reassembly.

The more serious accidents, leading to longer absence (31–90 days) relate most typically to materials, splinters and substances. Other significant accident causes include solid structures, i.e. doors, passageways, etc. Also in this group hand tools are one of the most typical injury causes. The typical accidents are, for example, a hand tool slipping during execution of the task, resulting into the face or hand injury. According to a study (Lind 2009), the most prevalent accident types in the case of fatal and severe non-fatal accidents are:

- Crushing or being trapped between components
- Person falling/jumping
- Accidents caused by falling/tumbling objects.

The accident contributing factors include:

- Dangerous working methods, including conscious risk taking (by workers)
- Defective hazard identification/unconscious risk taking (by workers)
- Non-use of PPE
- Working at a running process.

The chain of events leading to an accident can also involve latent factors that have an immediate effect, although they themselves do not cause accidents they are a contributing factor. Such factors include:

- Defective work instructions
- Defective safety equipment (machinery)
- The victim's lack of experience regarding the task involved
- Device failure
- Defective walking or working surface.

The following examples from real accident data describe some typical accident scenarios:

Example 1. A maintenance worker was welding when he got a spark in his eye.
Example 2. A worker was dismantling a system, when a heavy object fell onto his feet causing a minor injury.
Example 3. A worker was lifting an object, when he felt acute pain in his back.
Example 4. A worker was tightening a screw when the wrench slipped causing an injury.
Example 5. During a lifting operation, which was a part of a maintenance task, the load moved accidentally and the worker's hand was squeezed between a structure and the load. This caused some bone fractures and other injuries.

Of all accidents notified to insurance companies during 1999–2012 ($n = 12\ 088$) the most prevalent identified deviation relates to substances (leakage, puffing, etc.) (Fig. 2). Other prevalent, known deviations leading to accidents are e.g. losing control of a device or a tool, and a person falling, slipping and tripping.

It is significant that in a third of the cases the deviation is unknown. Together with the "other unlisted deviations," a major part of the deviations is unclear. The share of these groups may lead to misrepresentation of deviations and even complicate the effectiveness and focus of accident preventive activities. This indicates a clear need to improve accident reporting and investigation routines in maintenance-related accidents. The classification and reporting of accidents complies with Eurostat practices and is similar for all industries within the EU.

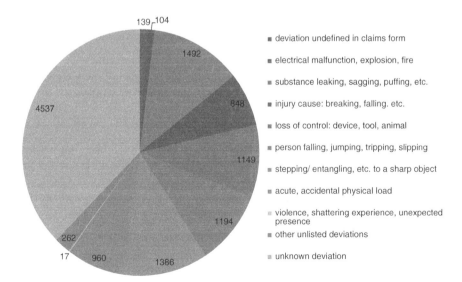

Fig. 2 Deviations leading to injuries in industrial maintenance in 1999–2012 (FAII 2014)

The accidents in industrial maintenance have certain specific characteristics. Table 1 summarizes the operations and deviations which have been the most prevalent at the moment of accident. Table 1 includes the combinations with one hundred or more reported accidents. The total number of accidents is 12 088.

The most typical accidents are falling, slipping, tripping and jumping, which is also the most typical occupational accident in Finland in general. Other typical cases are those involving sagging, leakage, eruption and puffing of a substance, leading to an eye injury. Such cases happen especially during welding and cutting. Other situations, specific for maintenance operations may occur while disassembling a technical system. In these cases, pipelines and other parts may still include substances or residues that can splatter during the work. According to the FAII's accident database, these kinds of accidents are very typical and may sometimes occur even though the worker is wearing safety goggles.

In addition, ergonomic problems (unexpected accidental physical load) are typical to maintenance operations as the operations can include lifting and moving heavy objects manually. Also the surrounding structures and other workplace-related issues can prevent the use of lifting equipment.

Table 1 Accidents in industrial maintenance: operations and deviations at the time of the accident (FAII 2014)

Operation/deviation	Person moving	Operating with manual hand tools	Operating a machine	Manual handling of objects	Moving a heavy object manually	Unknown
Falling, slipping, tripping, jumping	839			103	100	
Sagging, leakage, eruption, puffing (etc.) of a substance		682	285	258		
Loss of control: device, tool, animal, etc.		517	113	358		
Accident contributor: breakage, falling, etc.		196		300	123	
Injuring/stepping on a sharp object	330	354		446		
Unexpected accidental physical load	189			135	508	
Unknown						4 537
Total	1 586	2 013	604	1 702	951	4 537

4 Accident Prevention in Industrial Maintenance

The accident statistics and descriptions indicate that injury causes in maintenance operations vary greatly. The injury can be caused by energies and substances, the system with different parts and/or the surrounding operating environment, i.e. the structures and activities that do not relate directly to the maintenance task in question. Also the injured body parts vary in statistics. When exploring the maintenance-related hazards, the identification should be considered from three different viewpoints: (1) the task changes but the operating environment remains the same (e.g. outsourced factory maintenance with established customers sites), (2) the operating environment changes but the task remains the same (e.g. after-sales service), and (3) both vary (e.g. evening out temporary workload peaks between customer sites).

A specific feature for maintenance tasks is variability. Typically maintenance workers are constantly on customer's site, taking care of the most preventive and corrective maintenance tasks. In some cases, especially in the case of outsourced maintenance, the workers may change sites in order to even out the workload peaks. Task variability alone is a loss factor, as creating a safe but effective routine may be difficult and will take a long time. In the case of a new site with new tasks, cooperation with the customer is crucial. From the viewpoint of accident prevention, cooperation e.g. enhances risk assessment and task planning, as well as defining responsibilities between the maintenance workers and the operators on the customer's site.

Maintenance operations include typically both disassembly and reassembly of a technical system. This is a maintenance-specific feature, making maintenance different from many other tasks in industry. Disassembly is in particular risky, as, for example, the cause of a possible malfunction may be unknown, the failure mode misinterpreted, and/or the energies and substances may be hard to separate from the system. Moreover, disassembly and reassembly doubles the actual number of risks.

In order to reduce task-related risks in industrial maintenance effectively, the risk management activities must consider the maintenance-specific features. The accident preventive measures must tackle the hazards that may relate to the local workplace factors, organizational factors and/or the unsafe acts or practices. Moreover, the approach to risks can be distinguished from human- and technology-based failures and hazardous conditions. According to the findings from the previous studies, the most typical defects in safety and risk management relate to task planning, system design and task management.

5 Defective Task Safety Planning

Maintenance operations should be carefully planned and the related hazards identified in cooperation with the other operators on site. The risks should be assessed when there are new customer sites, and if/when there are new duties or tasks assigned for the maintenance workers. A brief, personal risk assessment before starting the maintenance work would help to identify any new and temporary risks that may have appeared since the last risk assessment.

6 Poor System Maintainability

This is a factor that is established during the system design phase. Also, later changes in the production/operating premises can create new problems. These problems lead to, among others, poor accessibility and reachability, as well as to direct injury risks. The tasks may also require more time and effort; if, for example, some structures and parts must be removed or dismantled before the actual maintenance operation. A specific group of problems relate to poor ergonomics. Difficulty in using lifting equipment and other devices that help to make the operation safe can lead to various ergonomics problems and injuries. Improving maintainability of an established system is difficult. Pay attention to work instructions, task allocation, resourcing, etc., so that the required time and resources are identified and planned realistically.

7 Defects in Management, Safety Culture and Customer Cooperation

Maintenance operations require specific site-related planning. Management should also pay attention to the local practices and customer cooperation on different sites. Maintenance operations should be in line with the safety practices, risk assessment and other local details. In addition, management should pay attention also to the safety culture existing between the maintenance workers on each site. There may be great variations between different sites. Maintenance operations may also be more susceptible to poor safety culture than some other operations in industry, due to e.g. the expectations towards quick problem-solving and inherent risks in industrial maintenance.

8 Lessons from Accidents: How to Identify Maintenance-Related Risks?

Maintenance-related risks can be approached from various viewpoints. These include, for example, task analyses, system safety analyses, risk assessments, etc. However, these may only help to partially identify the hazards, particularly if the focus is as such only on the system or on the activity. In order to manage maintenance-related risks systematically, the focus must be holistic, i.e. paying attention to both the system and the operation simultaneously. Understanding that major risks may be hidden in the system is crucial, as the risks may only be revealed or become activated when the system is being dismantled (energies, substances, etc.). Some risks can emerge only after the maintenance operation has been completed, i.e. in those circumstances in which the system is e.g. maintained or reassembled wrongly. Some focuses, while preparing and conducting maintenance-related risk assessments can be:

- Analysis preparation: is the aim to observe maintenance work and operating conditions in general, or is the aim to analyse one maintenance operation from the beginning to the end?
  - The prerequisite for the risk/task analysis method(s) to be chosen is that it must provide a holistic view of the risks relating to e.g. system safety, operating conditions and maintenance management. In the case of maintenance operations, analysis videos and/or photographs can be particularly useful, as they provide the possibility to revise details afterwards.
- Duties and practices on site: which tasks are allocated to maintenance workers and which tasks the customer (or somebody else) should take care of?
  - How the duties are defined and communicated to the maintenance workers?
- Other task details: is there maintenance worker on duty? Is she allowed to work alone? If yes, how is the safety and emergency planning carried out?
- Communications and cooperation between the maintenance workers and customer: what kind of cooperation is there between the maintenance workers and other operators on site? Do they have some common safety routines and updates (discussions, hazard identification, etc.)?
  - How are the maintenance task's orders delivered to the maintenance workers? What information do they get with the order: failure mode, safety details, urgency, etc.? What happens when the customer's workers notice the need for maintenance? Is there e.g. an established routine or order to conduct a task risk assessment on site before starting the actual maintenance operation?
- System safety design with regard to maintenance, maintainability: can the system be shut down easily and safely for maintenance work (N.B. emergency switches!)? Is the system's maintainability good? Does the design support easy and safe maintenance or add/increase the risks?
  - Do the typical tasks require plenty of preparation, e.g. is there a need to dismantle surrounding structures, etc., before starting the maintenance work? Is accessibility and reachability ok? N.B. these actions are also a part of maintenance work and they must be given consideration in the hazard identification/risk assessment.

- Do the maintenance operators have adequate tools and equipment for the work? Is appropriate lifting equipment available?
- How are the live parts, energies and (e.g.) chemicals switched off and separated from the maintained part of the system? N.B. steam, dust and compressed air.
  - Do the workers have personal safety locks and do they use the locks to make the system safe for maintenance?
  - How is the maintenance work marked/indicated for the other workers and operators on site?

9 Discussion and Conclusions

Maintenance activities can include a variety of tasks in almost any kind of industrial environment. In order to identify and manage the related risks, attention must be paid to human performance (alertness, skills, etc.), task planning (sequence, order, etc.) and technology (knowledge regarding hazards, structures, design, etc.). Below contribute the organizational aspects: are there enough resources, such as time and workers, to complete an operation safely and correctly?

A worker conducting a maintenance operation is an example of a socio-technical system, where both the technological system and the human are joint parts of the same entity. From the viewpoint of accident prevention and safety promotion, this reflects the need to manage the risks arising from the technology, together with the risks relating to a human. In the case of managing human-related risks, one has to consider both human actions as such, as well as organizational viewpoints.

There is a difference between scheduled, preventive maintenance and unscheduled repairs in regards to task and safety planning. The former is typically identified and even routine. This helps to plan the task details, including relevant safety measures, effectively. The latter is often troubleshooting, aiming to identify and locate the source of problem on the basis of hints and details. Possible misinterpretations may not only increase downtime, but can also cause some risks. This can be case when the system's condition with the related faults and effects is incorrectly identified. Thus, in risk management and safety planning specific attention should be paid to holistic risk management of unplanned repairs.

Accidents can have various underlying causes, mechanisms and chains of events. As maintenance is an exception from the normal, the accident causes and sources can be different in comparison to normal condition. This should also be reflected in the risk assessment and management procedures which should take into consideration maintenance operation sequence by sequence, taking into account factors and details in the activities, operating conditions and related technologies. However, accidents that happen during the maintenance operation can as such be similar to occupational accidents in general. Accidents can result from a single event or condition, but can also have their origins in small details that are tightly linked together (see Perrow 1999). In these cases, an accident results from a chain reaction as minor factors on e.g. the component level can together trigger a major event (see e.g. Heinrich 1959; Perrow 1999; Qureshi et al. 2007). In this perspective, component failure can be compared to human error, as they both can be the first initiative event, triggering the chain of events leading to an accident.

System resilience aims to manage and reduce tight couplings so that the system is more tolerant towards errors and faults (see e.g. Hollnagel et al. 2006). However, promoting safety during maintenance operations may not be possible or effective through resiliency. On the contrary, *the need for maintenance*, especially unscheduled repairs, may be reduced through resiliency, as a component failure (or an operator's human error) may not cause insuperable problems to the system. Safety, also in the case of maintenance, should thus be considered merely as a system property, not as a factor effected by one component or action only (see e.g. Marais et al. 2004).

As a conclusion, the maintenance-related risk assessment and management has to be holistic, so that factors relating to technology, human and organizational viewpoints are considered. The risk assessment and management procedures must also take into account the system's condition, surrounding activities and structures, as well as fault manifestation in practice. Depending on the scope, the analyses may also need to pay attention to post-maintenance system safety.

References

Dekker A (2006) A field guide to understanding human error. Ashgate Publishing, Dorchester

FAII (2014) Federation of accident insurance institutions in Finland. Accessed 14 April 2014

Heinrich HW (1959) Industrial accident prevention. A scientific approach. McGraw-Hill Book Company, New York

Hollnagel E, Woods DD, Leveson N (2006) Resilience engineering: concepts and precepts. Ashgate Publshing Limited, Surrey

Kumamoto H, Henley EJ (1996) Probabilistic risk assessment and management for engineers and scientists, 2nd edn. IEEE Press, New York

Lind S (2008) Types and sources of fatal and severe non-fatal accidents in industrial maintenance. Int J Ind Ergon 38:927–933

Lind S (2009) Accident sources in industrial maintenance. Proposals for identification, modelling and the management of accident risks. VTT Publications 710. Edita Prima Oy, Helsinki

Marais K, Dulac N, Leveson N (2004) Beyond normal accidents and high reliability organizations: the need for an alternative approach to safety in complex systems. http://sunnyday.mit.edu/papers/hro.pdf. Accessed 16 April 2014

Perrow C (1999) Normal accidents. Living with high-risk technologies. Princeton University Press, Princeton

Reason J (1997) Managing the risks of organizational accidents. Ashgate Publishing Limited, Aldershot

Reason J, Hobbs A (2003) Managing maintenance error. A practical guide. Ashgate Publishing Company, Aldershot

Statistics Finland (2014) PX-Web database: employment and working hours 1975–2012. http://pxweb2.stat.fi/Dialog/varval.asp?ma=080_vtp_tau_080&path=../database/StatFin/kan/vtp/&lang=3&multilang=fi

Qureshi ZH, Ashraf MA, Amer Y (2007) Modeling industrial safety: a sociotechnical systems perspective. In: Proceedings of IEEE international conference on industrial engineering and engineering management, Singapore, 2–4 Dec 2007, pp 1883–1187

Part II
Views on OSHM and Cases Towards Continuous Improvements

Chapter 5
Integrated Management Within a Finnish Industrial Network: Steel Mill Case of HSEQ Assessment Procedure

Maarit Koivupalo, Heidi Junno and Seppo Väyrynen

Abstract This chapter describes supplier's HSEQ management practices in general industrial context and in a shared workplace at a case company's premises. The case company is steel mill in Northern Finland. In addition, three supplying companies which operated in this shared workplace were selected to be a subject of research. HSEQ AP is a supplier's management procedure used by seven principal companies in Finland. The aim is to describe HSEQ AP in details, and the current situation of different supplier's HSEQ assessment practices in the case company. There were many different methods used for supplying companies' HSEQ management, and the principal company selected the main methods to be used: HSEQ AP and after-work HSEQ evaluation, in addition to safety requirements for service supplier form. Supplying companies' experiences of HSEQ AP were encouraging, but further development for HSEQ AP is needed. Safety indicators and the results of HSEQ AP were studied. Safety performance has improved in both the principal and supplying companies, but at the same time there have been many other safety activities in progress. Thus, it is not clear how much HSEQ AP has affected at the end to this positive development. Moreover, research for additional safety indicators is needed to prove more specifically the actual safety performance level, especially for supplying companies.

M. Koivupalo (✉)
Outokumpu Oyj, Terästie, 95490 Tornio, Finland
e-mail: maarit.koivupalo@outokumpu.com

H. Junno
Outokumpu Stainless Oy, Terästie, 95490 Tornio, Finland
e-mail: heidi.junno@outokumpu.com

S. Väyrynen
Faculty of Technology, Industrial Engineering and Management, University of Oulu,
P.O. Box 4610, 90014 Oulu, Finland
e-mail: Seppo.Vayrynen@oulu.fi

1 Introduction

Networking is a typical solution employed to help companies of different sizes manage in fast changing business environments. It is typical for employees from several companies, entrepreneurs and independent workers to work simultaneously in the same workplace. There has been a growing trend for principal companies to focus on their own special core business. This has led to the outsourcing of some of their usual work practices, mainly support processes. As a result, the number of shared workplaces (i.e. multiemployer worksite, shared worksite) is increasing.

The content of shared workplace in Finland is characterized by the employees of several employing companies, including all contractors, working at the same principal premises (Väyrynen et al. 2008). A company operating at the principal premises can be called inter alia a contractor, subcontractor, supplying company/network, provider, service delivering company/network, (external) supplier, service supplier or a partner. In addition, the principal company can be called for example the main company, a service purchaser, customer, principal employer or hosting organization. In this chapter, we use the terms *supplying company* and *principal company*.

Networking creates new HSEQ (health, safety, environment, quality) challenges at a national and global level. A single company, principal company, cannot necessarily choose which employees enter the site (Heikkilä et al. 2010). There can be permanent and non-permanent employees simultaneously at the same site and synchronizing their actions might be difficult. Safety challenges are for instance: accident prevention and safety at work (Nenonen 2011), safety risk management when personnel from several countries work at the same premises (Schubert and Dijkstra 2009) and inter-organizational communication flows (Heikkilä et al. 2010). Safety management procedures are rarely well suited to multiemployer worksites, and the need to reinforce safety management becomes even more important when service providers subcontract their operations further. The work may be performed by small companies, which do not have necessarily enough resources to manage safety activities adequately (Ylijoutsijärvi et al. 2001).

Requirements for managing HSEQ issues have to be met. Companies have also grown globally which increases HSEQ demands, and a renewed HSEQ management is required. One company cannot eliminate all HSEQ threats in shared workplace, and as a consequence there has to be a way to manage them. In a business environment, principal process industry companies require that their supplying networks provide them with high level and equal business solutions (and other related performance factors). At the same time, regulations emphasize the need for a high quality and equal management system. The management system includes the recording of environmental and work environment factors by companies of all sizes within the supply chain.

At the moment, there are several solutions available to manage HSEQ in shared workplaces. One solution for supplier's HSEQ management is *HSEQ AP* (*HSEQ Assessment Procedure*). HSEQ AP is used as a feasible method for meeting simultaneous business and regulatory needs and it is an assessment method which

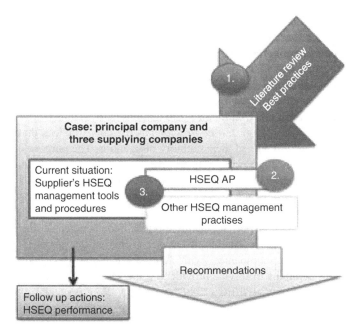

Fig. 1 Content and objects

can be used for evaluating supplying companies' HSEQ performance in shared workplaces.

The objects (Fig. 1) of this chapter are:

1. To give a description of HSEQ management in shared workplace as general industrial context
2. Describe the HSEQ AP in details
3. Find out the current situation of different supplier HSEQ assessment practices (tools and procedures) in the case company.

The final goal for this study is to obtain recommendations of how to improve principal company's contractor management practices and HSEQ AP. An important evidence of the HSEQ development is HSEQ performance indicator. Thus, also HSEQ indicators will be reviewed in this study.

2 Shared Workplaces and HSEQ Management

2.1 Legal Requirements in Shared Workplaces in Finland

In Finland, the legislation takes the cognizance of shared workplaces. A formal safety management system is not mandatory, but different approaches that are similar to a safety management system are required (Finlex 2002, 2006a, b).

Occupational Safety and Health Act (Finlex 2002) determines the duties and responsibilities in case of shared workplaces: '*each for their part and together in adequate mutual cooperation and by information ensures that their activities do not endanger the employees' safety and health*'. There is one company exercising the main authority (principal company), but health and safety management demands cooperation between each party.

Under the Act on the Contractor's Obligations and Liability when Work is Contracted Out (Finlex 2006b), *a contractor* has the duty to check that their contracting partner has discharged their statutory obligations. The purpose of the act is to promote fair competition between companies and compliance with terms of employment. This act is followed in the case company, whose role is comparable with a *main contractor*'s role.

Outsourcing is typical for Finnish workplaces in process industry. Maintenance, cleaning, security and transport are the most typically outsourced actions, and often a supplying company is responsible. It has been reported that about two-thirds of Finnish manufacturing companies (with at least ten employees) had transferred some of their operations to service providers in the early 2000s (Ali-Yrkkö 2007).

Council directive 89/391/EEC (1989) states that 'Where several undertakings share a work place, the employers shall cooperate in implementing the safety, health and occupational hygiene provisions and, taking into account the nature of the activities, shall coordinate their actions in matters of the protection and prevention of occupational risks, and shall inform one another and their respective workers and/or workers' representatives of these risks'.

The directive together with the legal requirements in Finland identifies similar actions needed in shared workplace as HSEQ AP seeks to solve (for more information see Sect. 6 in Chap. 2), e.g. highlighting the importance of cooperation, communication and prevention of occupational risks.

2.2 Safety in Shared Workplace

The value chain is a systematic approach to examining the development of competitive advantage. The chain consists of a series of activities that create and build value. They culminate in the total value delivered by an organization. To analyse the specific activities through which a firm can create a competitive advantage, it is useful to model the firm as a chain of value-creating activities. A value network is a business analysis perspective that describes social and technical resources within and between businesses (Porter 1985).

Networking has many advantages, for example synergies and economic benefits, reduced transaction costs, ability to concentrate on core skills or businesses, access to key technologies, and business risk sharing amongst partners (Hallikas et al. 2004; Heikkilä et al. 2010). On the other hand, many studies claim that outsourcing has negative impacts (c.f. Nenonen 2012):

- Coordination of different activities and implementation of safety measures is more difficult.
- Work is temporary and the environment changes rapidly.
- Cuts in staffing and reduction of qualified personnel.
- Service providers often have a limited overview of the customer company's operations, performed work tasks, special features and safety regulations.
- Service providers often operate in several principal companies' sites, where working practices, cultures and habits are different.

Accident prevention requires contributions from each individual person and operator in shared workplaces. When sending a worker to another employer's workplace, it requires a strong safety attitude and in-depth safety knowledge from the employee (Rantanen et al. 2007). For example, Lind (2009) has studied accident sources in maintenance operations, and has stated that the current increasing practice of subcontracting maintenance services creates new challenges: sites and tasks can vary according to the customer environment. In addition, there can be a wide variety of tasks which change according to the different operations assigned to the maintenance personnel.

If a company's supply chain does not obey comprehensive workplace safety practices, it can cause negative consequences, e.g. tragic loss of life or serious injury, higher insurance costs, financial and legal consequences, loss of corporate goodwill and difficulty to attract customers and recruit employees (Cantor 2008).

Nenonen (2012) concluded that many safety issues have already gained a great recognition in shared workplaces, e.g. communication and hazard identification. Simultaneously, management of some other safety areas has received limited attention, e.g. safety performance assessment. Likewise, implementation of safety cooperation between service providers and customers varied greatly. It is also important to take into account the different needs of permanent and temporary employees when developing a safety management system (Luria and Yagil 2009).

2.3 Integrated HSEQ Management

Safety management has a positive effect on the performance of safety, competitiveness and economic finance (Fernández-Muñiz et al. 2009). Many companies strive to maintain a high level of safety management, because they see it as a necessary competitive advantage. Business targets should be used to guide the company's safety efforts because they are a key factor for competitiveness, but also an important part of quality in business (Lanne 2007).

For many companies, it is not enough to settle for the minimum and so the targets are set higher. Setting targets higher than the minimum requirement provides a reliable picture of an HSEQ caring company that follows through socially, environmentally and economically sustainable principles. Organizations that choose to pay closer attention to safety, health and environmental factors would experience even more success on the way to continuous improvement (Mohammadfam et al. 2013). Socially responsible company invests to human capital and manages relationships with the social stakeholders (Montero et al. 2009).

Nenonen and Vasara (2013) concluded that companies in multiemployer worksites typically cooperated with their partners in safety management in training, orientation and guidance, flow of information, risk analysis, auditions and accident investigation. The implementation of cooperation was mainly successful, but strongly partner related. Parties at shared worksite should promote cooperation with implementation of safety activities to avoid overlapping safety operations, efficiently allocate remote resources for safety, strengthen commitment of all parties and ensure efficiency in safety.

Most Finnish principal companies in the field of forestry, chemical and metal process industries are taking care of corporate social responsibility (e.g. ISO 2010), sustainability policies (e.g. World Steel Association 2008) and practices. For example, World Steel Association (2014) states clearly that the most successful steel companies are also the safest.

Companies want to assure customers of the high quality of their products and services. This has led to the implementation of quality management systems and standards, such as ISO 9000. In addition, responsible organizations have to take care of the well-being of their employees, working environment, the impact of operations on the local community and the long-term effects of their products. This has led to the implementation of, e.g. EMAS, ISO 14000 and OHSAS 18001 international occupational health and safety management system specification (Levä 2003; Wilkinson and Dale 2007).

Separate management systems that cover quality, environmental, and safety and health issues have become too complex to manage effectively. For this reason, an integrated management system (IMS) has become of great interest to business Wilkinson and Dale (2007) have developed and integrated a management system model, which has the following definition: 'The part of the overall management system that includes the combined resources, processes and structures for planning, implementing, controlling, measuring, improving and auditing the combined quality, environment, and health and safety requirements of the organization'. As such, the level of integration of safety, environmental and quality management systems (or IMS) has become a popular research topic (e.g. Bernardo et al. 2009; Salomone 2008; Zeng et al. 2007).

Security is one essential contributor when considering contractor management practices in shared workplace. As well, security has been seen as one contributor for excellent business management (Tervonen 2010). Additionally, integration between quality, safety and ergonomics management has been investigated. Ergonomics seeks to design tools and tasks to be compatible with human capabilities and limitations with the purpose of providing work conditions that assures safety, health and well-being and efficiency (Dzissah et al. 2001).

Hamidi et al. (2012) list justifications to integrate HS, E and Q systems. The main reason is that there are many similarities, and integration reduces duplication and costs. They conclude that an integrated management system focuses on team work. Continuous improvement of quality, environmental aspects and health and safety can help to ensure that a company's leadership is committed to get on the continuous improvement journey towards sustainable development. According to Zutshi and Sohal (2005), the benefits are savings (time and cost), better utilization of resources and improved communication across the organization.

Sustainability and corporate social responsibility cover many similar aspects and principles as health, safety, environment and quality management and sustainable manufacturing. Zink (2014) concludes his research that there is a need for a different understanding of the overall performance of a company which can be seen in following the discussion about corporate social responsibility. This will generate new chances for human factors including the three dimensions of sustainability: economic, ecological and social.

2.4 HSEQ Performance and Indicators

Besides the principal company's own success and safety performance, the overall success depends on the performance in the entire network, including supplying companies' performance. Thus, HSEQ performance in shared workplace needs to be measured.

Safety performance indicators can be divided into leading and lagging indicators. *Lagging indicators* are gathered after losses have been incurred and cost assessments have been made (Grabowski et al. 2007). A growing number of safety professionals question the value of lagging indicators and argue that lagging indicators do not provide enough information or insight to effectively avoid future accidents (e.g. Grabowski et al. 2007; Mengolinim and Debarberis 2008; Hinze et al. 2013). There is a need to predict the future safety performance. *Leading indicators* address the need to predict and act before an unwanted event. Leading indicators measure the building blocks of the safety culture of a project or company (Hale 2009; Hinze et al. 2013).

Conversely, there is a debate going on of the division between leading and lagging safety indicators. Hinze et al. (2013) summarize parallel definitions (Table 1).

Recognizing signals and early warnings through the use of proactive safety indicators will reduce the risk of major accidents. There is no need for a discussion what is lead and what is lag, but there is a need to develop and implement useful

Table 1 Safety indicators and terminology (Hinze et al. 2013)

Leading indicators	Lagging indicators
Upstream indicators	Downstream indicators
Predictive indicators	Historical indicators
Heading indicators	Trailing indicators
Positive indicators	Negative indicators

indicators which can provide early warnings about potential major accidents (Øien et al. 2011). Dingsdag et al. (2008) stated that lead and lag indicators do not measure essential leadership attributes, communications and desired safety behaviours, and proposed to use instead Safety Effectiveness Indicators.

Defined safety indicators are an important role if a company seeks to improve its safety culture. A number of carefully selected leading indicators will probably provide the best predictive results (Hinze et al. 2013). As Morrow et al. (2014) conclude, safety culture is correlated with concurrent measures of safety performance and may be related to the future performance. The relationship between safety culture and safety performance is highly dependent on how and when both safety culture and safety performance are measured. Reiman and Pietikäinen (2012) summarize that safety management needs continuous focus on lagging indicators of past outcomes, including deficiencies and incidents, 'leading' indicators of current technical, organizational and human conditions and 'leading' indicators of technical, organizational and human functions that drive safety forward.

The lost time injury frequency rate (LTIFR) is the most commonly used indicator of HSE performance, and it is defined as the number of lost time injuries per one million hours of work. A lost time injury is an injury due to an accident at work, where the injured person does not return to work on the next shift. However, LTIFR has some deficiencies as a safety performance indicator. It is insensitive to the severity of the injuries, it is possible to manipulate the registration and classification of injuries, use of alternative job (i.e. restricted work), and for small companies LTIFR is fluctuating and sensitive to changes (Kjellén 2000).

Figure 2 shows LTIFR for World Steel Association member companies.

The overall trend is that most of the World Steel member companies have significantly improved their safety and health performance over recent years.

LTIFR for principal companies ($N = 5$) and supplying companies ($N = 4$) in Northern Finland, which have participated to HSEQ management projects is presented in Fig. 3.

LTIFR in these companies has decreased more rapidly than on average in Finland in the manufacturing industry.

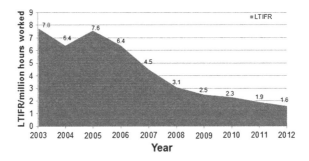

Fig. 2 Lost time injury frequency in World Steel Association companies (World Steel Association 2014)

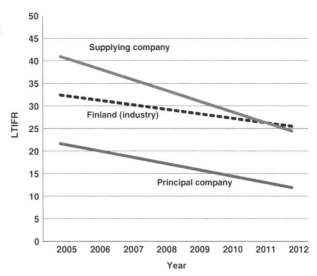

Fig. 3 Lost time injury frequency in converted to a linear regression trend (Väyrynen et al. 2014)

2.5 HSEQ Assessment Methods

Duijm et al. (2008) reviewed the present status of industrial HSE management in European Union member states. It was stated that there should be an investigation to determine how to most effectively manage HSE in the case of shared ownership and responsibility. Prime examples of this are in the form of industrial parks, leasing and outsourcing. HSEQ AP is a method that observes this feature and is one example of how to manage H, S, E and Q effectively.

There are several methods and tools that evaluate the performance of HS, E or Q. Some of these are developed to assess the performance of supplying companies while another have been developed to assess the principal companies' performance. Kjellén (2000) lists three different data collection methods: self-evaluation by a company rating team, rating by an independent assessor (third-party evaluations) and questionnaire to the workforce.

The international Safety Rating System (ISRS) uses standardized questions during data collection and fixed criteria in evaluating safety results. It has stimulated companies to develop their own rating system (Kjellén 2000). ISRS covers occupational health and safety management and sustainability-related issues including environmental, quality and security management, process safety management (ref. Seveso II Directive and OSHA 1910.119) and sustainability reporting (Global Reporting Initiative) (ISRS 2014).

There are several assessment tools for evaluating the safety (and health) performance of construction contractors. Ng et al. (2005) developed Safety Performance Evaluation (SPE), which was a safety performance assessment model used for evaluating construction contractors. El-Mashaleh et al. (2010) applied data envelopment analysis for evaluating construction contractors' safety performance.

Ai Lin Teo and Yean Yng Ling (2006) developed a method that can be used to assess the effectiveness of a construction firm's safety management system. The Construction Safety Index (SCI) can also be calculated by using this model.

Safety Checklist Contractors (SCC) is a standard that applies to the evaluation and certification of safety management systems (DNV 2010). Costella et al. (2009) introduced a method for assessing health and safety management systems (MAHS), which takes into account the principles of resilience engineering perspectives, which takes into account both operational and performance approaches. This method was tested at a factory that manufactures automobile exhaust systems. Chang and Liang (2009) developed a model which can be used to evaluate the performance of process safety management systems of paint manufacturing facilities, resulting in a Safety Index calculation.

Yang and McLean (2004) developed a template for assessing the corporate performance in EHS organizations. The Malcolm Baldrige Criteria for Performance Excellence is an application that can effectively be used when assessing and improving occupational safety and health management (Ketola et al. 2002). Occupational health, safety and environmental issues can be included in a balanced scorecard, which is an organizational performance measurement system (Mearns and Håvold 2003). The EFQM model (European Foundation for Quality management model) is based on total quality management (TQM) and is used as a self-evaluation and development tool. The model is used for the criteria in the European Quality Award and in most European national quality competitions (EFQM 2014).

A structural equation model of the construction safety culture was described by Chinda and Mohamed (2007). This model, as well as HSEQ AP, is based on the EFQM Excellence model. Mohammadfam et al. (2013) developed health, safety and environment excellence instrument, which was based also on EFQM, and more over to IPMA (International Project Management Award) and SCIM (The Safety Culture Improvement Matrix). Their tool measured HSE management system performance.

2.6 HSEQ Assessment Procedure

The Finnish Occupational Safety Card System was developed to improve occupational safety in a shared workplace (Väyrynen et al. 2008). The card is an *individual* certificate which indicates that an employee has basic knowledge about the cooperation and general hazards of the shared workplace, knowledge about the key principles and good practices, and is prepared to adopt workplace and job-specific orientation. The idea for a *company-specific* certification arose in the early 2000s after the individual certificate had been implemented. In addition to tools which are used to evaluate the organizational performance levels of HSEQ performance of supplying companies, it was the other important factor when process industry companies started to develop a HSEQ AP as a joint venture.

The object of the HSEQ AP is to ensure that supplying companies in shared workplaces possess enough knowledge about HSEQ requirements to operate in the sites of principal companies. It is also possible for the principal companies to evaluate the supplying companies' HSEQ performance (Niemelä et al. 2010a, b). HSEQ AP is used as a multicriteria performance indicator. Throughout the HSEQ AP, principal companies promoting contemporary responsibility and sustainability actions are included in their network. Supplying companies can be profiled according to HSEQ together or as one individual area (HS, E or Q) at a time. The maximum score in HSEQ AP is 750.

HSEQ AP is designed to (Hseq.fi 2014):

- Increase the productivity of a networked business.
- Improve the business skills in HSEQ matters.
- Encourage companies to develop systematic approach.
- Raise the level of management in companies.
- Help principal companies to select their own supplying companies.

The HSEQ AP as a process is described in details in Fig. 4.

The main phases are marked in numbers to Fig. 4 and the details are described as follows:

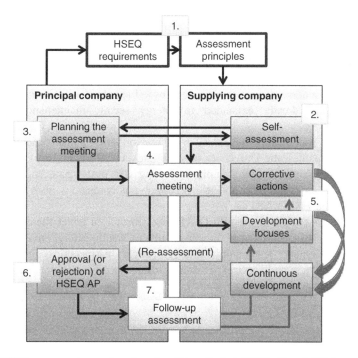

Fig. 4 HSEQ assessment procedure (based on Niemelä et al. 2010b; Hseq.fi 2014)

1. Main assessor sends a self-assessment to the supplying company before the actual HSEQ assessment meeting. At the same time, the main assessor provides detailed instructions what they expect from the supplier. This includes e.g. which documents should be available in the actual meeting, such as a copy of their safety policy and environmental strategy. Listing for 'Will my company get zero points from the HSEQ assessment' is also available (see Appendix).
2. Self-assessment is made using the same HSEQ questions as in the actual HSEQ meeting.
3. Main assessor plans and arranges the meeting. If required, the main assessor seeks clarification from supplying company before the meeting.
4. In actual HSEQ meeting evaluation team (consisting of the main assessor and up to three assessors from different principal companies) visits the supplying company to perform the HSEQ assessment.
5. A report of the HSEQ assessment is produced and it is given to the supplying company. The supplying company has a predetermined period (approximately 3 months) to define the corrective actions for development focuses and any deviations if noted.
6. Main assessor decides if the identified development focuses and deviations are corrected properly. Results are reported to Hseq.fi register (HSEQ.FI database).
7. After 3 years, a follow-up assessment is performed.

There are about ten thousand employees together in seven principal companies that are involved in HSEQ AP. As May 2014, 95 HSEQ assessments have been completed and 18 assessments are currently running. The size of evaluated supplying company varied from very small to international corporation. From process industry, HSEQ AP seeks to find new users e.g. from port operators (Turunen 2014) and information and communication technology (Lakkala 2014). Group of principal companies has already expanded to energy sector, maintenance and tyre manufacturer.

HSEQ AP was mainly developed by five Northern Finnish process industry principal companies, the University of Oulu and POHTO (The Institute for Management and Technological Training). During the past years, members have developed to its current form (Fig. 5). Supplying companies also took part in developing the HSEQ AP.

The questions used in the HSEQ AP assessment tool cover the same area as EFQM model (EFQM 2014) in the perspective of health, safety, environment and quality management: leadership; strategy; people; partnerships and resources; processes; products and services; customer results; people results; society results; and key results (c.f. Appendix).

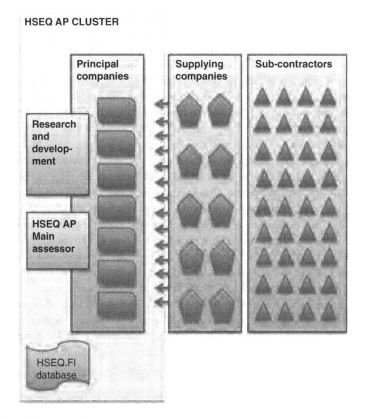

Fig. 5 HSEQ AP network

3 Materials and Methods

Empirical section (next chapter) can be divided into three main sections, which review supplier's HSEQ management practices from different aspects:

(1) Supplier HSEQ management history in the case company
(2) Supplier HSEQ management and the current situation in the case company
(3) Supplying companies' views related to HSEQ AP.

Both principal company and supplying companies were located in Northern Finland. The principal company had about 2,300 employees on site, which was a subject of research in this chapter.

The main method for historical part (1) was interview for one key employee who had been working in the principal company since the beginning of 2000s. The interview was semi-structured. Interviewee's responsibilities were related to safety (and later environment) during his career in the case company and supplier's HSEQ management was major part of it. With the help of him and archives, the historical path of supplier's HSEQ management was formed.

For the current situation (2), the main method was questionnaire but also document review was used. The results of questionnaire were gathered into an internal project report. The questionnaire was received by 22 recipients, and it was sent to 23 (response rate was 96 %). The respondents represented different departments around the principal company including procurement, production, maintenance and mill services, logistics, HSEQ departments, research centre and port operations.

Three representatives from supplying companies were interviewed for the third part (3). They were responsible of HSEQ issues in the supplying companies. It was known in advance that they had the widest knowledge about supplier's HSEQ management practices related to their own companies in the premises of the principal company. The supplying companies offered maintenance, repair and installation services for process industry and the number of employees was about 40, 60 and 80. Semi-structured interview questions were provided to the interviewees before the interview.

Empirical section also analysed HSEQ performance indicators, using the data from principal company's archives and HSEQ AP results (with the maximum scores of 750). The main indicator was chosen to be LTIFR, which was defined similarly with Kjellén (2000) (as described in HSEQ performance and indicators section). LTIFR is defined as the number of lost time injuries per one million hours of work.

4 Results

4.1 History

History of supplier's HSEQ management in the principal company's premises started to develop systematically in the beginning of 2000s. At the same time, research projects were implemented (for more information about the path of projects towards integrated management system in process industry compare to Väyrynen et al. 2012).

- 2004–2006 basic requirements were defined for suppliers. Health and safety were in focus and the requirements were based on literature and good practices. The first version of after-work HSEQ evaluation was drafted.
- 2006–2007 a piloting model of HSEQ AP was developed and the first pilot assessments were implemented. In addition to health and safety, environmental and quality aspects were combined to the procedure. The first HSEQ assessors were trained during this time period.
- 2007–2009 cooperation between principal and supplying companies and HSEQ AP developed further. Moreover, assessment tool and assessor training had enhancements. After-work HSEQ evaluation was implemented and systematic HSEQ assessments with web-based assessment tool were started.

- 2010–2012 the work continued and HSEQ AP was developed further as a goal to achieve national usage. Supplier's experiences were reviewed and the assessments continued using the web-based assessment tool (HSEQ.FI database). Main assessor's role changed from principal companies to third party, which was among other functions specialized to certification.

4.2 Current Situation

4.2.1 Principal Company

By the end of 2012, HSEQ AP was developed and used at some level in the case company. After-work HSEQ evaluation was in use as well, but not in a greater extent. At that time, it was noticed that supplier's HSEQ management in the premises of principal company was fragmented and included too many different methods. Inside the principal company, different assessment methods were used by different organizations without necessarily knowing each other's actions. Also, the reasons for using some methods were not clear for everyone. Some instructions advised to use a particular method, but the instructions were not well-known, assigned nor followed by a wider group. The best methods were not exploited with the full potential and supplier's HSEQ performance level was not followed systematically using the same methods by different parties.

The case company decided to start a project to organize and systematize its service supplier's HSEQ management. One reason for this was that many suppliers worked in the shared workplace and it was important to be able to choose partners whose HSEQ performance was at an adequate level. In addition, supplying companies' LTIFR was obviously higher compared to the own personnel's equivalent frequency (see Fig. 8).

During the project, many methods were discovered, and the total number was 16–19 (depending on the method of calculation). Some of the methods were used only once and some of them were not in use at the time the questionnaire was made. All methods did not cover health and safety issues in-depth, but concentrated more to quality (HS and E aspects were treated slightly). On the contrary, some of the methods were focussing only on HS.

It is possible to categorize the methods and practices into four groups:

1. Requirements, e.g. legal requirements and internal requirements for supplier's occupation safety.
2. Assessment and self-assessment methods, e.g. HSEQ AP (Hseq.fi 2014) and SafetyTen (Anttonen and Pääkkönen 2010).
3. Steering meetings, e.g. supplier's safety meetings.
4. Evaluation of the performance at work, e.g. after-work HSEQ evaluation.

One goal for the project was to identify overlapping methods and practices, and the results showed that there were some overlapping practices. However, some of the methods were not in wide usage and they were used by different departments. Thus, overlapping practices were not common although different departments practiced similar kinds of activities. Some of the assessment methods were used for choosing suppliers, but there was not any formal and common practice for the selection.

During the project, opinions about HSEQ AP from the representatives in principal company were received (translated from Finnish):

- Method gives good practices to current partner rather than operates as a supplier selection criteria. It is not possible to rank suppliers based on HSEQ AP.
- HSEQ AP is almost complete system, which has been developed during the past years. It offers own internet page, you can participate to training and there are instructions and descriptions available. There is also one database for the documents.
- Using the system saves time for both: principal company and supplier company, because various principal companies do not need to do many assessments separately. It is possible to exploit assessment results if some principal company has participated, it does not need to have own participation always.
- HSEQ AP is too expensive for suppliers.
- HSEQ AP is not very agile. The process is inflexible and it is not possible to apply it in any way (on the other hand, it is the same for all standardized methods).

4.2.2 Supplying Companies

Opinions regarding HSEQ AP from supplying companies were quite similar to those in 2010 (Niemelä et al. 2010b), although the HSEQ AP practice had changed during the past 3 years. One fundamental difference is that the main assessor's responsibility changed from principal companies to the third party (specialized to certification).

Suppliers' representatives agreed that assessor group was aggregated properly and they thought that it is definitely important to have representative (or if possible several of them) from principal companies. In addition, it was important to have assessors who had different backgrounds (e.g. from safety department and from procurement) to exploit the full potential of HSEQ AP. Good practices are the best that HSEQ AP have to offer, and that is another reason why the assessor group must be diverse. One interviewee had a concern regarding the main assessor's qualification, especially if there are many main assessors.

All suppliers found one defect in practical arrangements: they were not instructed clearly enough before the assessment meeting. It was not automatically obvious what was expected from them, and they had to ask additional questions and instructions. Another defect was that follow up for deviations and development targets was not implemented clearly and systematically enough. The representatives from supplying companies were not sure if the assessment was finalized or was something still expected from them.

Every interviewee thought that the requirements and results were realistic and that the duration of assessment meeting was suitable. Latva-Ranta et al. (2012) discovered that HSEQ AP self-assessment results are similar to the actual assessment, and it supports the perception of realistic results. Practical arrangements were functioning well and there was time for visiting the site during the assessment meeting, which was perceived to be very important. The assessment tool was functioning well and it was easy to use in everyone's opinion.

Safety performance indicators showed a positive trend for each supplying company (compare to Fig. 6 and Fig. 7). When asked how significant a role HSEQ AP had in this development, all interviewees thought that it supported the process, but the main influence was with other actions which the company had made towards a safer workplace. HSEQ AP functioned mainly as a marketing tool giving a positive image about the company. In addition, it offered examples of valuable good practices related to HSEQ. They stated that it did not play an important role when gaining new customers, not did it have a direct influence on the supplying company's HS performance or financial importance.

All interviewees hoped that HSEQ AP would provide a more transparent procedure, and for example well-performing companies would be listed publicly in the internet (at the moment, they are listed based on a identification number). This comment also supports the other view of how well-known HSEQ AP is. All representatives thought that in Northern Finland it is known, but only in bigger companies. Even inside the principal company, which was a subject of research in this study, HSEQ AP is not well known among all.

All interviewees agreed that cooperation with the principal company had improved after they participated to HSEQ AP. Nevertheless, the significance of HSEQ AP results was not clear for the interviewees. They were not sure how the result would affect the end result when the principal company selects suppliers. Two of the representatives commented that it is not fair to require participation from some suppliers and then choose other suppliers who have not participated to the process. They questioned what is the importance when they participated to HSEQ AP, and how do they benefit afterwards? They also thought that the assessment was too expensive, because it did not affect the choice of supplier.

Benefits for the shared workplace and its HSEQ performance were not direct in any interviewee's opinion. They thought that it has some effect regarding a long-term development, but that the safety culture does not change rapidly. Companies that have not invested yet strongly to HSEQ, would benefit the most and for these companies, HSEQ AP would have the most to offer. As a summary, all representatives from supplying companies thought that HSEQ AP has some positive affect to HSEQ performance and practices in shared workplaces.

4.3 HSEQ Performance

Figure 6 illustrates HSEQ AP results (maximum 750) for supplying companies ($N = 11$), which had participated twice to the assessment.

All supplying companies which had participated to two assessments got better results on the second assessment round. Average increase was 44 % when the second assessment was compared to the first assessment. Suppliers A, B and C (Fig. 6) participated to the interviews in this chapter. On average, these three companies increased their scores more (63 %) than on average (45 %).

LTIFR for the supplying companies which participated to this research is illustrated in Fig. 7.

LTIFR development trend has been decreasing for all of these three companies. It is remarkable that the higher the LTIFR was in 2008 the faster development has been during the past 6 years.

Figure 8 shows LTIFR progress in the shared workplace, in the premises of the case company during the past 10 years.

Trend has been decreasing for both own personnel and suppliers. During the last 6 years, suppliers have had about 1/3 of the working hours compared to the supplying companies' own personnel working hours in the shared workplace.

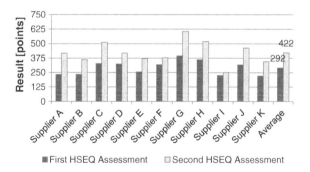

Fig. 6 Comparison between the first and the second HSEQ AP results

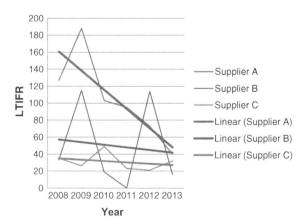

Fig. 7 LTIFR progress in the supplying companies

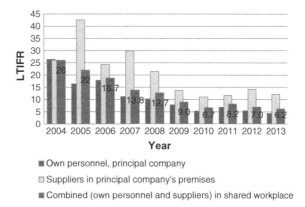

Fig. 8 LTIFR in shared workplace

5 Discussion

5.1 Summary of the Current Situation

HSEQ issues are a daily topic in shared workplace in the case company. It is not possible always to choose which employees enter the site, especially when subcontracting is used. The principal company has demanded to improve HSEQ practises in shared workplace and due to that different supplier's HSEQ management methods were developed and applied. However, practices were fragmented and formal standard procedure did not exist.

In the beginning of the project, the number of different assessment methods was high, and there was a clear need to make a decision and selection. This procedure was created, the internal instructions were specified and the methods were chosen. The main assessment method was chosen to be HSEQ AP. One goal for HSEQ AP is to help principal companies to confirm supplier's adequate HSEQ level. However, before the project it was not yet used extensively for that purpose. Representatives from principal company did not see HSEQ AP as a method which could rank suppliers. After-work HSEQ evaluation was decided to be used after the work is completed, among other detailed internal criteria, which was based on the contract type, duration and value.

Before the project, both methods were not in wide usage, and the goal is next to increase the usage. After some time has passed, there will be a review how the usage has developed. For every contract which requires that principal company has checked that their contracting partner has discharged their statutory obligations (according to the Act on the Contractor's Obligations and Liability when Works is Contracted Out, see Sect. 1 in Chap. 2), safety requirements for service supplier form needs to be filled in and approved. The form is valid for 2 years. HSEQ AP is required from those supplying companies which have longer than one year contract. For minor contracts, extended safety requirements for service supplier form will be used if HSEQ AP is not required. Now when the instructions are specified and HSEQ AP is selected to be the main method, it will also increase the importance for suppliers to participate to HSEQ AP.

5.2 Development Requirements for HSEQ AP

Experiences about HSEQ AP revealed quite similar observations than previous studies (c.f. Niemelä et al. 2010b). Although, few years have passed the process is almost the same. HSEQ AP is functioning well in basic level, but some improvements and remarks would make it even better method for supplier's HSEQ management:

1. Ensure that suppliers have good instructions for HSEQ AP: they know how to prepare and what to expect from the assessment.
2. When making the first contact with the supplying company to suggest the participation to HSEQ AP, the purpose and significance needs to be stated clearly, so that suppliers know what advantage they will get if they participate to HSEQ AP.
3. Evaluation team needs to be heterogeneous and have enough participants also in the future. This is one of the strengths HSEQ AP has and taking care that it will remain the same ensures the success of HSEQ AP during longer period.
4. After the assessment meeting, HSEQ AP as a process needs to be finalized properly. It includes taking care that deviations and development focuses are finalized formally and sending of the final report.
5. When there are many main assessors in Finland, requirements for assessors' qualification needs to be strict. Main assessors must be qualified to achieve actual results from assessments.
6. HSEQ AP in its current format is not widely well known and it needs more visibility. Also, the system should be more transparent.
7. Supplying companies which have participated to HSEQ AP should have access to the best practises database. At the moment, best practises are collected during HSEQ AP meeting to HSEQ.FI database, but suppliers do not have access to them. In the assessment meeting, the assessors may not think to inform about a particular practise if they have not participated to the assessment meeting covering this practice.

Most of these (1–5) are easy to improve just by ensuring that current procedure is followed precisely. Hence, the process how to improve these development focuses already exists. However, the last two (6 and 7) would require adjustments to the current practices and the system.

HSEQ AP can be used to share good HSEQ practices, and that is the main purpose how it is perceived at the moment in the case company. HSEQ AP has been able to improve HSEQ skills and management in those companies which have participated to HSEQ AP, e.g. by adding knowledge. Another remarkable advantage is the saving of time and resources, and consequently productivity. This advantage could be highlighted even more than it has been. Companies who have participated to HSEQ network have been able to show evidence of positive safety performance development. One advantage that has not been exploited yet entirely in the case company is to help principal companies to confirm supplier's adequate HSEQ level. HSEQ AP has been used for this purpose at some level, but not to a great extent. After this step is executed in a wider extent, suppliers will see more value for participating to HSEQ AP.

5.3 HSEQ Indicators

The scores in HSEQ AP show that all participants have increased their HSEQ AP results significantly after the first assessment. The supplying companies which participated to this research had even greater increase in their scores. However, the sample of three companies is small, and it is not possible to conclude if these three companies have actually improved the level more than on average, and what is the reason for that. Still, it is clear that these three companies have taken HSEQ issues very seriously and invested to be better in this field during the past few years.

Because overall success depends on the performance in the entire network, suppliers' safety performance has been followed by the principal company for many years. Safety performance indicators (in this case LTIFR) for principal and supplying companies showed that safety in shared workplace has improved during the last decade. Also, perceptions with larger amount of companies and longer time scale which have participated to HSEQ AP support this conclusion. Time scale for supplying companies' LTIFR in this chapter was too short to draw any high-level conclusions besides that the trend is positive.

Global trend for LTIFR in steel producing companies is also decreasing, and clearly LTIFR is in lower level compared to the four companies in this chapter. However, it is not possible to compare these figures directly, because it is not known what lies behind these numbers. For example, it is not known what the response rate is, which organizations have responded to the survey, in what extent they use restrictive work, how their national legislation takes into account the absence due to workplace injuries, and how they have determined LTI that will be included to their statistics (there might be some injuries that occurred in workplace that will not be approved). Nevertheless, it is possible to say that the trend is positive also in this sample, but more research would be needed to draw conclusions and to be able to compare these statistics.

Furthermore, criticism related to LTIFR as an indicator needs to be noted, e.g. for small companies it is fluctuating and sensitive to changes. If the real HSEQ performance level seeks to be determined, also other lagging indicators and leading indicators would need to be reviewed. This chapter provided preliminary review for HSEQ performance and clearly further research is needed.

6 Conclusions

Although many safety issues have gained recognition in the shared workplaces and many tools and methods have been developed (especially for contractors in construction sites), there are still many HSEQ and security threats related to supplier's management in global business environment. The whole supply chain needs to be considered and a comprehensive safety culture in shared workplace needs to be established. If company seeks to perform higher than average in the terms of HSEQ, they must ensure of corporate social responsibility and sustainable processes. HSEQ AP can be one method to support that purpose.

From the principal company's perspective, adding usage for HSEQ AP and after-work HSEQ evaluation will be the next steps towards a safer shared workplace. It must be ensured that the methods are in wide usage and every relevant employee has an access to the databases and knows how to use them. HSEQ approach can be an essential method in value creation, both in intra- and inter-organizational sense. The key objective of an enterprise resource planning (ERP) system is to integrate information and processes. The ERP software applications help to improve operational efficiency and productivity of business processes of a company or value network. HSEQ comprises information and processes to be integrated within ERP. The actual result how HSEQ AP and after-work HSEQ evaluation will succeed as tools in supplier's HSEQ management will be seen after few years.

Appendix

WILL MY COMPANY GET ZERO POINTS FROM THE HSEQ ASSESSMENT? (Hseq.fi 2014)

Look over the list of claims below. If all of them are true you will receive at least 250 points out of 750 points.

HS—Health and safety issues

- The company is aware of the competence and know-how of the staff (documentation exists). Know-how is improved by courses and internal training.
- Managers and supervisors have received occupational safety training targeted to managers which includes the responsibilities of occupational safety.
- Working hours of the employees are monitored. A tracking system is in use.
- Occupational health care has been arranged according to legislation, work place inspections and health check-ups have been performed.
- A competent occupational safety manager is nominated. The manager supports and develops the cooperation in occupational safety and health matters supporting the line organization. The occupational safety manager has been notified to the Occupational safety personnel register.

- Occupational safety representative and deputy safety representative have been nominated.
- A systematic course of action of developing occupational safety exists. It includes the goals, methods and responsibilities of occupational health and safety as well as competences and description of occupational health care.
- Work-related harms and risk factors (including the ones related to the environment) have been identified and the risks they cause have been assessed.
- Accidents are investigated and documented.
- There is a procedure for processing the hazardous incidents.
- Supplier aims to clarify if its own network of suppliers has the criteria and procedures to fulfil the demand stated in legislation.

E—Environment issues

- Waste sorting has been instructed and trained and containers for different sections of waste exist. Sorting know-how is included in the orientation.
- The company occasionally keeps track of changes in the legislation (e.g. checks the legislation related to an order).
- The company has mapped its environmental effects and acts to decrease them (e.g. acts required in legislation).
- A person has been named to be in charge of the maintaining the list of chemicals and getting the material safety data sheets.
- Environmental deviations are registered and dealt with. A list of actions is created if needed.
- The company takes responsibility of environmental issues and informs the interest groups.
- There is at least one indicator for societal results and it is locally compared. (Theses, internships, work hours of externals, environmental goals set by the customer, development of the amounts of waste, functionality of waste sorting or amounts of harmful substances).
- Functionality of waste sorting is checked (e.g. from the waste containers), maximum of one error per container is accepted.
- Different sections of waste are measured, some of them have goals and the trend is improving. Information for comparison is available.

Q—Quality issues

- There is a board or team of management and proof of its activity exist (records).
- The company's methods related to quality, environmental issues, occupational health and safety have been informed to the personnel.
- These aspects of operating are measured regularly and the essential data are informed to the personnel.
- Strategic goals, aims and actions have been documented in the company's plans. The personnel and other interest groups have been informed about the strategy.
- HSEQ-related goals, aims and actions have also been documented in plans.
- Orientation procedures exist and they can be proved to be applied.

- Satisfaction is measured (e.g. work satisfaction inquiries, development discussions).
- The personnel are rewarded for good performance.
- The company has evidence of systematically developing its own network of suppliers and partners in multiple fields of HSEQ.
- The working space is functional. Everyone takes care of the tidiness of their own post. The level of cleanliness is adequate.
- Work equipment is task specific. Spare equipment/parts exist. Traceability of measurement has been taken care of.
- Methods and technologies of work are functional and are developed in case of any problems come up. Problems have been analysed.
- Processes have been identified and are documented with HSEQ aspects taken into account.
- The responsibilities of development and planning have been set and HSEQ aspects are included. The customer needs for example has been used as source information.
- Orders are recorded in an appropriate tracking system.
- Customer feedback is collected, summarized and informed to the interest groups.
- The entire system of processes has been documented (e.g. quality handbook) and commonly known evaluation methods of operations (e.g. SafetyTen) have been implemented.
- Customer satisfaction is measured.
- Replies are given to all feedback and their reasons are taken care of.
- Barometers are placed into active use and comparison material is available. Goals are set and the results are mediocre at the minimum.
- Frequency of incident reports is monitored and a goal is set. The frequency has improved in past years.
- Employee satisfaction is measured repetitively.
- Accident frequency is monitored. A target is set for it and improvement has been made during past years.
- Frequency of personnel absence is monitored. A target is set for it and improvement has been made during past years.
- Percentage of sick leave is monitored and a goal is set.
- Suggestive initiatives are received from personnel and teams.
- The realization of training investments is monitored, goals are set and comparison data exists (e.g. number of training days).
- Economic indicators, such as revenue, income, gearing ratio, revenue per employee, etc., are in active use and comparison data are available. Goals have been set and the results are mediocre at the minimum.
- Other indicators that describe the activity well, such as comparisons of productivity, level of invoicing, comparisons of cost efficiency, realization of strategic indicators, etc., are in active use and comparison data are available. Goals have been set and the results are mediocre at the minimum.

References

Ai Lin Teo E, Yean Yng Ling F (2006) Developing a model to measure the effectiveness of safety management systems of construction sites. Build Environ 41(11):1584–1592

Ali-Yrkkö J (2007) Outsourcing in Finnish manufacturing—does industry matter? ETLA, The Research Institute of the Finnish Economy, Finland (in Finnish, English abstract). http://www.etla.fi/wp-content/uploads/2012/09/dp1070.pdf. Accessed 8 Aug 2014

Anttonen H, Pääkkönen R (2010) Risk assessment in Finland: theory and practice. Saf Health Work 1(1):1–10

Bernardo M, Casadesus M, Karapetrovic S, Heras I (2009) How integrated are environmental, quality and other standardized management systems? An empirical study. J Clean Prod 17(8):742–750

Cantor DE (2008) Workplace safety in the supply chain: a review of the literature and call for research. Int J Logist Manage 19(1):65–83

Chang JI, Liang CL (2009) Performance evaluation of process safety management systems of paint manufacturing facilities. J Loss Prev Process Ind 22(4):398–402

Chinda T, Mohamed S (2007) Structural equation model of construction safety culture. Eng Constr Archit Manage 15(2):114–131

Costella MF, Saurin TA, de Macedo Guimarães LB (2009) A method for assessing health and safety management systems from the resilience engineering perspective. Saf Sci 47(8):1056–1067

Council Directive 89/391/EEC (1989) On the introduction of measures to encourage improvements in the safety and health of workers at work. http://eur-lex.europa.eu/LexUriServ/LexUriServ.do?uri=CELEX:31989L0391:EN:NOT. Accessed 11 Aug 2014

Dingsdag DP, Biggs HC, Cipolla D (2008) Safety effectiveness indicators (SEI's): measuring construction industry safety performance. In: Thomas J, Piekkala-Fletcher A (eds) Third international conference of the Cooperative Research Centre (CRC) for construction innovation—clients driving innovation: benefiting from innovation, cold coast, Australia, 12–14 Mar 2008

DNV (Det Norske Veritas) (2010) SCC (Safety Checklist for Contractors). http://www.dnvba.com/US/certification/management-systems/Health-and-Safety/Pages/SCC.aspx. Accessed 2 May 2010

Duijm NJ, Fiévez C, Gerbec M, Hauptmanns U, Konstandinidou M (2008) Management of health safety and environment in process industry. Saf Sci 46(6):908–920

Dzissah JS, Karwowski W, Yang Y-N (2001) Integration of quality, ergonomics, and safety management systems. In: Karwowski W (ed) International encyclopedia of ergonomics and human factors, vol 2. Taylor & Francis, London, pp 1129–1135

EFQM (2014) The EFQM excellence model. http://www.efqm.org/the-efqm-excellence-model. Accessed 2 May 2014

El-Mashaleh MS, Rababeh SM, Hyari KH (2010) Utilizing data envelopment analysis to benchmark safety performance of construction contractors. Int J Project Manage 28(1):61–67

Fernández-Muñiz B, Montes-Peón JM, Vázquez-Ordás CJ (2009) Relation between occupational safety management and firm performance. Saf Sci 47(7):980–991

Finlex (2002) Occupational safety and health act 738/2002. http://www.finlex.fi/en/laki/kaannokset/2002/en20020738. Accessed 19 Mar 2010

Finlex (2006a) Act on occupational safety and health enforcement and cooperation on occupational safety and health at workplaces 44/2006. http://www.finlex.fi/en/laki/kaannokset/2006/en20060044. Accessed 19 Mar 2010

Finlex (2006b) Act on the contractor's obligations and liability when work is contracted out 1233/2006. http://www.finlex.fi/en/laki/kaannokset/2006/en20061233. Accessed 19 Mar 2010

Grabowski M, Ayyalasomayajula P, Merrick J, McCafferty D (2007) Accident precursors and safety nets: leading indicators of tanker operations safety. Marit Policy Manage 34(5):405–425

Hale A (2009) Why safety performance indicators? Saf Sci 47(4):479–480

Hallikas J, Karvonen I, Pulkkinen U, Virolainen V-M, Tuominen M (2004) Risk management processes in supplier networks. Int J Prod Econ 90(1):47–58

Hamidi N, Omidvari M, Meftahi M (2012) The effect of integrated management system on safety and productivity indices: case study; Iranian cement industries. Saf Sci 50(5):1180–1189

Heikkilä A-M, Malmén Y, Nissilä M, Kortelainen H (2010) Challenges in risk management in multi-company industrial parks. Saf Sci 48(4):430–435

Hinze J, Thurman S, Wehle A (2013) Leading indicators of construction safety performance. Saf Sci 5(1):23–28

Hseq.fi (2014) HSEQ assessment procedure. https://www.hseq.fi/index.php?. Accessed 2 May 2014

ISO (2010) ISO 26000 project overview. International Organization for Standardization, Genève

ISRS (2014) ISRS—for the health of your business. http://www.dnvba.com/Global/sustainability/management-practices/Pages/isrs.aspx. Accessed 13 Aug 2014

Ketola J-M, Liuhamo M, Mattila M (2002) Application of performance-excellence criteria to improvement of occupational safety and health performance. Hum Factors Ergon Manuf 12(4):407–426

Kjellén U (2000) Prevention of accidents through experience feedback. Taylor & Francis, London

Lakkala J (2014) Developing Supplier Auditing Network for ICT Sector. Lapland University of Applied Sciences, Finland

Lanne M (2007) Co-operation in corporate security management. VTT Publications 632, Espoo

Latva-Ranta J, Väyrynen S, Koivupalo M (2012) HSEQ-palvelutoimittaja-arvioinnin käytön laajentamisen ja vaikutusten seuranta. Project Reports of Work Science 32, Department of Industrial Engineering and Management, University of Oulu, Oulu

Levä K (2003) The functionality of safety management systems: strengths and areas for improvement at six different types of major-hazard installation. Safety Technology Authority, Helsinki

Lind S (2009) Accident sources in industrial maintenance operations. Proposals for identification, modelling and management of accident risks. VTT Publications 710, Espoo

Luria G, Yagil D (2009) Safety perception referents of permanent and temporary employees: safety climate boundaries in the industrial workplace. Accid Anal Prev 42(5):1423–1430

Mearns K, Håvold JI (2003) Occupational health and safety and the balanced scorecard. TQM Mag 15(6):408–423

Mengolinim A, Debarberis L (2008) Effectiveness evaluation methodology for safety processes to enhance organizational culture in hazardous installations. J Hazard Mater 155(1–2):243–252

Mohammadfam I, Saraji GN, Kianfar A, Mahmoudi S (2013) Developing the health, safety and environment excellence instrument. Iran J Environ Health Sci Eng 10:7

Montero MJ, Araque RA, Rey JM (2009) Occupational health and safety in the framework of corporate social responsibility. Saf Sci 47(10):1440–1445

Morrow SL, Koves GK, Barnes VE (2014) Exploring the relationship between safety culture and safety performance in U.S. nuclear power operations. Saf Sci 69:14–28

Ng S, Cheng KP, Skitmore RM (2005) A framework for evaluating the safety performance of construction contractors. Build Environ 40(10):1347–1355

Nenonen S (2011) Fatal workplace accidents in outsourced operations in the manufacturing industry. Saf Sci 49(10):1394–1403

Nenonen S (2012) Implementation of safety management in outsourced services in the manufacturing industry. Tampere University of Technology, Tampere

Nenonen S, Vasara J (2013) Safety management in multiemployer worksites in the manufacturing industry: opinions on co-operation and problems encountered. Int J Occup Saf Ergon 19(2):167–183

Niemelä M, Latva-Ranta J, Väyrynen S (2010a) Development of one Finnish model for HSEQ management—HSEQ AP. In: Labart L, Pääkkönen T (eds) Towards better work and well-being, programme and abstracts. Finnish Institute of Occupational Health, Helsinki, Finland, 10–12 Feb 2010, p 68

Niemelä M, Latva-Ranta J, Ollanketo A, Väyrynen S (2010b) TUOLATU—Uudenlaisten HSEQ-edellytysten havainnointi yritysverkostossa, TUOLATU-Observing new HSEQ perquisites in company network. Project Reports of Work Science 28, Department of Industrial Engineering and Management, University of Oulu, Oulu

Øien K, Utne IB, Herrera IA (2011) Building safety indicators: part 1—theoretical foundation. Saf Sci 49(2):148–161

Porter ME (1985) Competitive advantage: creating and sustaining superior performance. The Free Press, New York

Rantanen E, Lappalainen J, Mäkelä T, Piispanen P, Sauni S (2007) Accidents at shared work places—lessons to be learned. Työ ja ihminen 21:364–379

Reiman T, Pietikäinen E (2012) Leading indicators of system safety—monitoring and driving the organizational safety potential. Saf Sci 50(10):1993–2000

Salomone R (2008) Integrated management systems: experiences in Italian organizations. J Clean Prod 16(16):1786–1806

Schubert U, Dijkstra JJ (2009) Working safely with foreign contractors and personnel. Saf Sci 47 (6):786–793

Tervonen P (2010) Integrated ESSQ management. As a part of excellent operational and business management—a framework, integration and maturity. University of Oulu, Oulu

Turunen H (2014) Analysis of health, safety, environmental and quality management in a network of port operators. University of Oulu, Oulu

Väyrynen S, Hoikkala S, Ketola L, Latva-Ranta J (2008) Finnish occupational safety card system: special training intervention and its preliminary effects. Int J Technol Hum Interact 4(1):15–34

Väyrynen S, Koivupalo M, Latva-Ranta J (2012) A 15-year development path of actions towards an integrated management system: description, evaluation and safety effects within the process industry network in Finland. Int J Strateg Eng Asset Manage 1(1):3–32

Väyrynen S, Jounila H, Latva-Ranta J (2014) HSEQ assessment procedure for supplying industrial network: a tool for implementing sustainability and responsible work systems into SMES. In: Ahram T, Karwowski W, Marek T (eds) Proceedings of the 5th international conference on applied human factors and ergonomics AHFE 2014, Kraków, Poland, 19–23 July 2014

Wilkinson G, Dale BG (2007) Integrated management systems. In: Dale BG, van der Wiele T, van Iwaarden J (eds) Managing quality, 5th edn. Blackwell Publishing, Oxford, pp 310–335

World Steel Association (2008) 2008 sustainability report of the world steel industry. World Steel Association, Brussels

World Steel Association (2014) Safety and health. http://www.worldsteel.org/steel-by-topic/sustainable-steel/social/safety-and-health.html. Accessed 30 Sept 2014

Yang Y, McLean R (2004) A template for assessing corporate performance: benchmarking EHS organizations. Environ Qual Manage 13(3):11–23

Ylijoutsijärvi P, Latva-Ranta J, Luomanen J, Sulasalmi M, Vesala A (2001) Työturvallisuustoiminnan kehittäminen teollisuuden alihankinnoissa (TYKTA)—loppuraportti. In: Ylijoutsijärvi P (ed) Työturvallisuustoiminnan kehittäminen teollisuuden alihankinnoissa—TYKTA. Project Report, University of Oulu, Oulu

Zeng SX, Shi JJ, Lou GX (2007) A synergetic model for implementing an integrated management system: an empirical study in China. J Clean Prod 15(18):1760–1767

Zink KJ (2014) Designing sustainable work systems: the need for a systems approach. Appl Ergon 45(1):126–132

Zutshi A, Sohal A (2005) Integrated management system—the experiences of three Australian organisations. J Manuf Tech Manage 16(2):211–232

Chapter 6
Introducing a Scenario of a Seaport's HSEQ Framework: Review and a Case in Northern Finland

Hanna Turunen, Seppo Väyrynen and Ulla Lehtinen

Abstract This paper addresses health, safety, environment and quality (HSEQ) issues, specifically on the HSEQ Assessment Procedure (AP) of a Finnish industrial network. The need for and practical application possibilities of using the HSEQ AP in the management of a port is addressed in detail. The port is situated within the northern Barents Region. The study concentrates on the information acquired from previous implementations of this AP in the supply network of the paper, chemical and steel companies (process industry) and the existing criteria and practices of evaluations using this audit-style approach. The empirical interview part of this study was conducted mainly in Northern Finland. Some information and impressions on HSEQ were also gathered during a visit to Russia's Murmansk Commercial Seaport, in the largest part of the Barents Region. The paper has its document-based and interview-based parts. The former consists of a thorough review of scientific and industrial documents and the latter refers to hearing about HSEQ issues in the field from the interviewed representatives of a maritime company network (a case). The combination of these approaches provides us with recommendations and scenarios for further consideration.

Keywords Seaports · Maritime · Health · Safety · Environment · Quality · Management system · Assessment · Suppliers · Value network

---

H. Turunen · S. Väyrynen (✉)
Faculty of Technology, Industrial Engineering and Management,
University of Oulu, P.O. Box 4610, 90014 Oulu, Finland
e-mail: Seppo.Vayrynen@oulu.fi

H. Turunen
e-mail: hanna.e.turunen@gmail.com

U. Lehtinen
Oulu Business School, University of Oulu, P.O. Box 4600, 90014 Oulu, Finland
e-mail: ulla.lehtinen@oulu.fi

1 Introduction

This paper aims to review health, safety, environment and quality (HSEQ) management concepts, especially Finnish ones, in the context of seaports. The unique direct views from the field perspective comprise study material from a company network within the maritime sector. The empirical part of this study on HSEQ issues was interview-based and conducted mainly in Finland in the Port of Kemi and in Oulu. Some information and impressions on HSEQ were also gathered in Russia in Murmansk Commercial Seaport. During the end of summer through the fall of 2012, the document-based research was conducted, and the field phase was carried out in 2013.

The study aims to clarify the status of HSEQ conditions and management issues of the logistical sea transport industry, especially in seaports. HSEQ standards are upheld through the passing and enforcement of regulations in developed societies and through voluntary efforts in operations and businesses. A generic example of the former is legislation on occupational health and safety (cf, Ministry of Social Affairs and Health 2013), and of the latter, quality awards and responsible management models (cf, EFQM 2013; ISO 26000 2010). Specifically regarding seaports, good practice models include the Brisbane health and safety guides with emphasis on contracting companies (Health and Safety 2013) and European Union (EU)-based guides for management and sustainability of seaports in Europe (ESPO 2012). The stakeholders, i.e. personnel, customers, suppliers, owners, society and the Globe, are also waiting for better performance in HSEQ and sustainability issues (cf, ISRS 2014; Worldsteel 2013).

HSEQ integrates HS issues with other aspects. A lot of material exists from many countries addressing specifically HS issues in the seaport industry and transport by water: one good example is the material from the United Kingdom (UK) (Managing 2002; Safety in Docks 2014). On the other hand, a definition of accidents by the UK Health and Safety Executive supports an integrated approach. By this definition, an accident is "any unplanned event that results in injury or ill health of people, or damage or loss to property, plant, materials or the environment or a loss of business opportunity" (Hugnes and Ferrett 2003). We think this definition corresponds with HSEQ's integrated approach. Harms-Ringdahl (2013) addresses the issue in this way: "Quite often, auditing also includes safety, health and environmental aspects, since the management of these is similar and sometimes also integrated. Variation concerning which elements should be included is large."

The HSEQ Assessment Procedure is a practical tool developed to evaluate health, safety, environment and quality management in the supply network of the process and manufacturing industry (Väyrynen et al. 2012). It offers companies an evaluation performed by an outsider and thus, a bias-free reference for companies in the process of comparing and choosing suppliers. These suppliers also have an opportunity to ask for an HSEQ Assessment Procedure (later referred to as HSEQ AP) to be carried out for their network's suppliers.

The HSEQ AP is a relatively new research subject as it is still a new practice, even in the process industry, though the HSEQ AP has a 20-year-long development history in the process industry (cf, Väyrynen et al. 2012). The research on the HSEQ AP as a tool in logistic operations is lacking because, thus far, the HSEQ AP has only been used in the process industry. The aim of this study is to show the adaptability of the HSEQ AP to ports and to describe the possible changes its implementation would bring.

This study addresses the information acquired from previous implementations of the HSEQ AP and the existing criteria and practices of evaluations using an audit-style approach. Harms-Ringdahl (2013) defines an audit as a procedure that examines a company's management system to see whether it conforms to some kind of (external) norm.

The most relevant EU regulations regarding employment, working conditions and safety in waterborne transport are presented by the European Agency for Safety and Health at Work (2011). In the transport industry, including waterborne transport, occupational safety statistics reflect a relatively high number of accidents. According to the European Statistics on Accidents at Work (ESAW), the incidence rate of non-fatal occupational accidents (i.e. cases per 100,000 workers of non-fatal occupational accidents (more than 3 days lost)) decreased in the total working population between 1994 and 2006. This was true for all three transport subsectors (land transport (road, train and pipelines), water transport and air transport). However, while in the total working population the downward trend in the incidence rate was constant, in transport the trend was somewhat different. For instance, in water transport, there was a reduction in the rate of accidents until 2003, after which the incidence rate of non-fatal occupational accidents increased through 2006. In 2006, the average incidence rate of fatal accidents was 3.5 in the total working population, while in the land transport sector it was 14.7 and in the water transport sector it was 15.3 (European Agency for Safety and Health at Work 2011).

The special background for this study is the increasing interest in the northern Barents Region (including Russia) and the Barents Sea. This study is also part of the EU-funded Barents Logistics 2 project that aims to facilitate cooperation and business between the Barents countries. Collaboration is stimulated through scientific research, business partnerships and general studies of the area. The arctic Barents Region has been enjoying increased attention in the media and in scientific studies due to its vast natural resources and potential for economic growth. This potential predicts growth in industry and business, which leads to a growing demand for logistical solutions in the coming years. Most probably more research and development on logistics is needed, too. One topic of research and development will be HSEQ management. The HSEQ AP might be one of the tools to guarantee sustainable logistics in this huge, very special and vulnerable region.

The research questions for this study are:

1. How does the HSEQ AP work and what are its main principles?
2. Can the HSEQ AP be used to create solutions with additional value to the port industry?

3. What kind of changes would the HSEQ AP bring to the Port of Kemi?
4. What can be said about the adaptability of the HSEQ AP to the port industry according to this study?

1.1 The Barents Region and the Port of Kemi

The Barents Region is a vast area centred on the Barents Sea. The borders of the area vary but it includes areas in the northernmost parts of Norway, Sweden, Finland and Russia. The population is scarce; the area covers about 1.75 million sq km and is populated by about 5.23 million people. In addition to the nationals of the four previously mentioned countries, there are three indigenous peoples: the Nenets, the Saami and the Veps. The largest cities of the area are Murmansk in Russia (pop. 300,000), Archangelsk in Russia (pop. 350,000), Rovaniemi in Finland (pop. 61,000), Oulu in Finland (pop. 190,000), Umeå in Sweden (pop. 110,000), Luleå Umeå in Sweden (pop. 75,000), Tromsø in Norway (pop. 69,000) and Bodø Tromsø in Norway (pop. 48,000) (barentsinfo.org).

Although the area consists of four different countries whose relations have not always been amicable, a good deal of cooperation exists between the countries. The Barents Euro-Arctic Council (BEAC) was officially established in 1993. According to the official declaration, it is meant to "serve as a forum for considering bilateral and multilateral cooperation in the fields of economy, trade, science and technology, tourism, infrastructure, educational and cultural exchange, as well as projects particularly aimed at improving the situation of indigenous peoples in the North" (The Barents Euro-Arctic Region 1993). The cooperation happens at two levels: intergovernmental and interregional. The BEAC includes Denmark, Finland, Iceland, Norway, Russia, Sweden and the European Commission.

The Barents Region is famous and strategically important for its rich natural resources such as forests, minerals, fish, oil and gas. The climate is quite harsh. The history of the area is interesting. The Russian city of Murmansk is a nuclear military base that had huge strategic significance during the Cold War. Then the borders of the former Soviet Union were tightly guarded. The reformation of Russia has been a stimulated interaction in the north. Due to the remoteness of the area, there is much to do in regard to developing logistics and infrastructure and protecting the fragile climate. In the future, the area is likely to have even greater strategic value due to the reasons presented next (Zimmerbauer 2013).

As the climate warming causes the Arctic sea ice to thin, the Northern Sea Route can be used more intensively. So far the Route has been open for 2 months per year but researchers predict that that the time window will increase. The Northern Sea Route offers a route from Europe to the Pacific Ocean that is 14,000 km shorter compared to the current one, allowing reduced shipping time and associated costs. The ports in the Barents area are likely to become busier as more ships start to pass through. Obstacles still exist, as the coastline between Barents and the Bering Strait is practically uninhabited, which means no stopovers for restocking or technical aid are available there. The challenges presented by the harsh climate are also significant (Verny and Grigentin 2009; Borgerson 2008).

The EU has shown interest in the arctic and sub-arctic regions of northern Europe by including them in the Northern Dimension, a policy framework aimed at fostering increased dialogue and concrete cooperation between northwest Russia, Kaliningrad, the Baltic and the Barents Sea, the Arctic and Sub-Arctic areas. The Northern Dimension is also intended to strengthen stability and economic integration in the region, while keeping in mind the sensitive area of ecologic sustainability. It has four different partnerships that focus on health and social wellbeing, culture, environment and transport and logistics (The Barents Euro-Arctic Council 2013).

Regions and ports in northern Finland, especially Kemi, think about their possibilities for certain key roles in the development and utilization of the Barents Region and Sea. This scenario of the lively Northern Sea Route is linked to the logistics needs of Finland and the possible role of Finland and the Baltic Sea as a Finnish and European channel to and from the Barents Region and Sea.

1.2 Seaports in General and the Port of Kemi Ltd.

Ports can be defined as the contact point between sea traffic and land transport. This definition can include the equipment and buildings of ports and the services offered in the area. Often goods also need to be stored before leaving the port, either by land or by sea. It should be noted that the word "transport" in this definition can refer to transport of both people and goods. The vocabulary used in EU and Finnish legislation uses the term port facility, which refers to the area where the actual interaction between a ship and a port happens. This includes docks, waiting slots and fairways in and out of the port. One port can include more than one port facility (Karvonen and Tikkala 2004).

As ports offer many things to customers, they have many operators working within them. The goods meant for shipment might, for example, have to be transported to the port, stored and loaded on a ship, and have the necessary paperwork completed. These tasks require the efforts of many people for their smooth and safe completion. The network of these operators and businesses and their interest groups is called a port community. There are private and municipal operators in a port as well as representatives of the state, in the form of customs and border guards (Karvonen and Tikkala 2004). A sketch of port community participants is presented in Fig. 1.

The Port of Kemi (2014) is fully owned by the City of Kemi, although it was incorporated in the beginning of 2012 by changing its organisation and name to the Port of Kemi Ltd. The Port of Kemi Ltd. owns the land occupied by the piers and other port structures in its area. It does not own the buildings or equipment within the area, only the land. The Port of Kemi Ltd. does not handle cargo. The services it

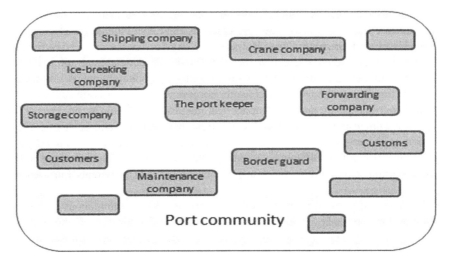

Fig. 1 A sketch of the likely participants of a port community (modified from Karvonen and Tikkala 2004)

offers to ships are towing and icebreaking in the area surrounding the docks. Renting land to companies that work on the piers is the biggest part of the port operations. There are five piers in the port with a maximum draught of 10 m, but plans for an expansion have been made.

The Port of Kemi has undergone some changes since its incorporation in 2012. Among other changes, as the Port of Kemi Ltd., the company has revised its job descriptions and roles and updated its organizational chart. Its general attitude towards cost-effectiveness has become more positive. Its history of being part of the city still affects operations. For example, occupational health and safety operations are the responsibility of the City of Kemi. Occupational health care is still provided by the city's communal health services as no changes were made during the process of incorporation.

The Port of Kemi is a relatively small port in comparison with other ports around the world. In 2011, 420 ships from 20 countries visited the Port. The Port is fenced and guarded by the port supervisors so that no unknown traffic is allowed in or out. Until the International Ship and Port Facility Security (ISPS) code was put into effect by International Maritime Organization (IMO), the Port was open for anyone to visit.

The role of the Port of Kemi Ltd. in the Port is similar to a landlord. It encourages free competition and thereby avoids the establishment of a monopoly. As the landowner, the Port of Kemi Ltd. is required to supervise and maintain the Port's facilities, including the dock and the piers and all other parts of the infrastructure. The Port of Kemi Ltd. also has the right to collect rental fees and port charges from companies that use its facilities. If cargo is kept in the Port for a notably long time, rent can be charged for this as well. The Port of Kemi Ltd. also has the authority to demand that ships in its area move or to leave if necessary.

The Port of Kemi, like all corresponding ones, can be described as a common (shared) workplace which has its own legislation. When more than one operator works simultaneously (or consecutively) at a workplace and there is a designated main authority in the area, the workplace is called a shared workplace. In Finland, a shared workplace has its own legislation which prescribes specific roles for the main authority and the other operators. All employers and employees of the workplace are required to communicate sufficiently in order to avoid any dangers or threats to the health and safety of employees. The main authority is responsible for appropriately informing all employers and employees of the possible harm and dangers in the area, providing guidelines and directives for the work and the workplace and measures of fire prevention, first aid and evacuation, as well as the people responsible for those services. At the same time, all the employers and employees in the area are responsible for informing each other of the harm and dangers their work might cause. The main authority, the Port of Kemi Ltd., is also responsible for consolidating the work performed in the area, traffic control, the order and the tidiness and general planning of the workplace, as well as general healthiness and safety. Cooperation between the members of the workplace regarding occupational health and safety is to be initiated by the main authority.

An independent actor in a shared workplace must be aware of the possible harm and risks he causes, must have the necessary permits and licenses in effect, must use appropriate equipment and tools safely and perform the necessary periodical and implementation inspections. An independent actor must use appropriate personal protection equipment if necessary or required and follow regulations on handling hazardous substances. In addition, he must follow any safety instructions issued by the main authority. The shared workplace model is analogous to an umbrella, where the main operator is the umbrella offering safety to the companies in its shade, making sure that every company fits and does not harm others by its actions (Fig. 2).

The Port of Kemi Ltd. is a small workplace: it only employs about 35 people. Its job descriptions vary from normal office work to more physical and outdoor port officer tasks. The possible harm related to office work comes mainly from static muscular tension that is commonly caused by using a computer. Disorder and untidiness indoors may cause falls and thus injuries; slippery conditions due to cold weather and inadequate gritting outdoors may cause falls as well. For port officers, the job includes operating outdoors, which exposes them to cold weather and its attendant problems. Operations on the piers and ships carry with them the hazards of falling and traffic-related accidents. The constant movement in the port resembles that of a busy factory and so the risks are about the same. Being hit by a car, train, or moving machinery comprises a potential cause of a severe damage.

The Port of Kemi Ltd. does not produce much waste, as a large part of its work is carried out in an office. As the main authority, it has the responsibility to organize the handling of waste. Port officers monitor the amounts and kinds of waste that the whole Port produces and how well it is sorted. The company is also responsible for

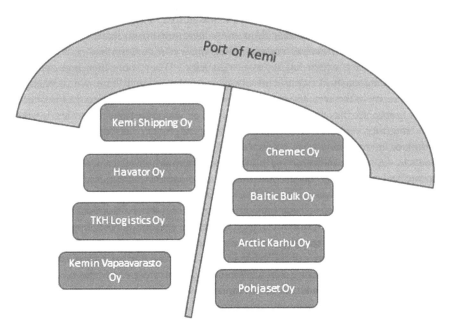

Fig. 2 The main operator offers safety and order to the companies in the shared workplace

obtaining an environmental permit for the Port due to the likely expansion of the Port in the near future.

The Port of Kemi is geographically comprised of three separate port areas: Ajos Port, Veitsiluoto Port and a separate oil port. This study focuses on Ajos, as the oil port is focussed only on oil transport and Veitsiluoto mostly only serves the Veitsiluoto paper plant owned by Stora Enso. Ajos Port is located in the northern part of the Bothnian Bay, close to the border between Finland and Sweden. Because of its location, Ajos offers port service and a route to northern Finland, the northern parts of Scandinavia and to the Russian parts of the Barents Region. The port is mainly focused on export.

The simplified stream of goods in a port is presented in Fig. 3. The figure does not take into consideration the required paperwork or inspections, just the basic logistic functions in a port. A port is a link between land and sea transport. According to Lun et al. (2010), a port is an essential part of the national immobile logistics infrastructure as well as a dealer of commercial infrastructure and a business gateway.

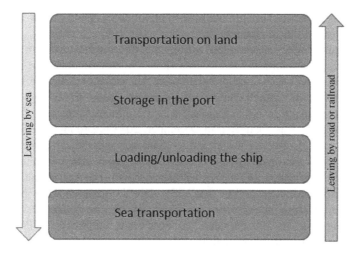

Fig. 3 A sketch of the stream of goods in a port

2 Method and Materials of the Empirical Study

The empirical part of this study was carried out using qualitative research; more specifically, through interviews used to answer the research questions. The interviewees represented the three case study companies operating in the Port of Kemi. The current status of HSEQ of the Port of Kemi was outlined in the interviews and possible gaps or problems in implementing the HSEQ AP are analysed. Then possible changes brought by an imaginary implementation were estimated and evaluated. This study comprised only a hypothetical draft of the changes to come; the actual realization of those changes is left to the Port of Kemi to decide.

Case studies can be used for the purposes of organizational and management studies. A case study is an in-depth investigation that uses qualitative research methods such as direct observation and interviews to gather information for analysis. Findings can then possibly be used when creating a generalization or theories (Yin 1989). This study was based on semi-structured interviews. This method lets the interviewees express their opinions and perspectives (Hirsjärvi et al. 2007; Sinclair 2005). The empirical research method included predominantly qualitative interviews rather than focus group-style sessions (Sinclair 2005). The three companies that were studied, by interviews with key person(s) from each, are all operators of the Port area: the Port of Kemi, Kemi Shipping Oy and Havator. Those three companies were chosen based on their key roles in port operations and because together they provide a small sample of the supplier chain inside the port. The Port of Kemi Ltd. is the port official and main actor (as it owns the land), Kemi Shipping Oy is the main shipping company and supplier of stevedoring services of different kinds, and Havator is responsible for most of the crane and lifting operations in the port. These interviews also provided an opportunity to observe the relations between the companies.

Additionally, a key representative from the If insurance company, operating mainly in Nordic Countries, was also interviewed in order to obtain the perspective of a member of an interest group. A company that provides insurance is of course interested in how an insured company manages its operations.

The first company to be interviewed was the Port of Kemi Ltd. A short preliminary interview was conducted in December, 2012 regarding the HSEQ issues. The goal was to clarify the roles of and relations between the companies operating in the Port. The main interview was performed in March, 2013. The interviewees present were the finance and marketing manager and the head of transport, both very well informed about HSEQ issues. So were the representatives of the other case study companies, too.

Kemi Shipping (2014) is the main partner in cooperation with the Port of Kemi Ltd. as it does most of the stevedoring, forwarding, clearance and storage-related tasks in the Port. It is the company that has the most to do with the cargo while a ship is in the Port. It owns the storage buildings and the lifting equipment. For larger scale or specific lifting, it cooperates with Havator. It also provides repair work for the ships if needed. Kemi Shipping is also in contact with other companies in addition to the Port of Kemi Ltd. because it offers various services. It cooperates with Havator and with the land transport companies, port officials, customs, ship owners and other companies needed to make the cargo flow efficiently through the Port.

The law requires a shared workplace practices, thereby ensuring smooth transport in the Port. Kemi Shipping has its own environmental permit for its repair shop; in addition, the company operates under the environmental permit issued to the Port of Kemi Ltd. This is an example on how closely the companies are bound together. As Kemi Shipping is in charge of most of the internal traffic in the Port, it is vital that the rules and regulations of the Port of Kemi Ltd. are followed.

Kemi Shipping's employee job descriptions vary greatly as do the ones of the Port of Kemi Ltd. Some of the work is normal office paperwork, some is loading and unloading ships, trucks and trains using heavy machinery, and some are manual repair work. In addition to the general safety instruction offered by the Port of Kemi Ltd, Kemi Shipping gives new employees a full occupational instruction including safety issues.

Havator (2014) acts as a crane operator in the Port of Kemi. Havator is a local company that was founded in Tornio. Its early operations included trading grains, lumber and war-related scrap metal and operating mills and sawmills. The business converted entirely to construction work in the 1950s. Company headquarters are still located in Tornio. Today Havator has four divisions: Havator Cranes, Havator Montage, Havator Transport and Havator Cranes. Havator Cranes is focussed on crane services for both long- and short-term projects. Havator Montage specializes in installing and assembling large scale concrete elements and steel structures. Havator Transport offers specialized transport services for extra-large and extra heavy objects. It also offers jacking and skidding services (Qvist 2010). Polar Lift used to be a separate company offering harbour crane services but it is now fully merged with Havator Oy.

Havator specializes in project work and often performs it in cooperation with other companies. A recent project entailed mounting windmills at the coastline of the Gulf of Bothnia. Havator operates in Finland, Sweden, Norway, Baltic countries and Russia, which links it to the Barents Region and makes it relevant for the present study. It has become the biggest crane service provider in Nordic countries and one of the biggest in Europe, thanks to its recent acquisitions in Finland and Sweden.

Havator's opinion is that it has a good HSEQ reputation. At the time of this study, the company had already passed the HSEQ AP in 646/2011. Havator has also been granted ISO 9001 (ISO 2008), ISO 14 001 and OHSAS 18 001 certification. As Havator operates in several countries, the certificates provide proof of good quality, safety management and environmentally friendly operations on an international scale. The phrase "safety first" has become a slogan for the company. As it operates in the Arctic region, it also has valuable operating experience in a climatically challenging environment.

3 Empirical Results

3.1 Interviews of Representatives of the Port of Kemi

Both interviewees from the Port of Kemi Ltd. played an important role in HSEQ-related issues and had a lot of knowledge and skills of them. As previously mentioned, the conversation was open, rather than being confined to precise questions and direct answers.

The Port of Kemi Ltd. has been granted ISO 9 001 (ISO 2008), ISO 14 001 (ISO 2004) and OHSAS 18001 (BSI 2007) certification which is a good base for defining the HSEQ-related status for an organisation.

The Port of Kemi Ltd. documents all trainings and courses as part of its quality documentation. Holding the Finnish Occupational Safety Card (cf, Väyrynen et al. 2008; Safety Card 2014) is seen as a minimum requirement for all employees in relation to health and safety issues. Typically, there is no single specific educational background required from workers of the Port. The clerical employees and supervisors do not have so a diverse background. They often have more formal educational backgrounds. The working hours are recorded with an enterprise resource planning system module specifically designed by for port environment.

Due to its history of being part of the organisation of the city of Kemi, the Port buys its occupational healthcare services from the city. The services include the health check-ups and workplace inspections required by Finnish law. Employees are also free to get a health check-up sooner than required. The Port is still included in the occupational safety organization of the city of Kemi and both the occupational safety representative of the employees and the manager of the employer are appointed via the city's organization. The Port of Kemi is interested in the occupational health and safety of the people of the other employers in the port but requires no compulsory reporting, as information is voluntarily shared. As the main operator of a shared workplace, the Port of Kemi organizes basic orientation for any new employer in the area and training in case there are essential changes in the area. Once a year, regular and temporary workers of the Port are reminded of the safety and security issues and traffic rules of the area. All new operators undergo training regarding ISPS code and its effects and access to the area.

The environmental issues are taken care of. Comprehensive environmental mapping was completed in the port around 10 years ago as a part of construction to make the Port a deep-water port. A few years later, another mapping was completed for the extension of the Port and in connection with this. The Port of Kemi was granted an environmental permit in 2005. The permit requires careful examination of the environmental aspects of operating. As a part of the permit requirements, the Port of Kemi has to provide and update a waste management plan and provide the docked ships with a waste disposal guide. Although the Port of Kemi itself employs about 35 people, it does not produce much waste. Mostly the waste is produced by the companies operating in the Port or the docked ships leaving the Port of Kemi. According to the permit, these companies and ship operators are responsible for monitoring and reporting their waste. All new operators of the port are informed about the permit restrictions and regulations. Agreements are signed, since all the operators in the Port are operating under the environmental permit of the Port of Kemi Ltd.

For the Port officers, waste monitoring is part of their daily routine. In addition, the head of maintenance operations (who has been more thoroughly trained in the screenings of waste) goes through the waste containers once in a while and provides a report of his findings. The Finnish law on waste control (646/646/2011) was renewed in 2011 and required some waste disposal updates in the Port. Once every 3 months, the provider of waste disposal services compiles a report of all the waste. By measuring the amount of waste, the Port of Kemi can track whether they have reached the goals set in their plans. Boat discharges are measured and estimated with a discharge count system used in all Finnish ports, created by the VTT Technical Research Centre of Finland and Finnish Port Association. Particle measurements are taken every 5 years and the rain water system is inspected once a year.

Some chemicals are shipped through the Port. The Port of Kemi receives a material safety data sheet for each of the substances. Each sheet provides information on the properties, associated risks and safe ways to use and store the substance. The responsibility for these chemicals lies with the following key persons: onboard a ship, it is the captain; on land, the head of each post, e.g., the head of maintenance operations is responsible. When a new hazardous substance is to be transported via the port, it must be negotiated with the authorities that decide whether or not to grant the permit. Due to the new copper and nickel enrichments being transported via the port, the Finnish Centre for Economic Development,

Transport and the Environment (ELY) demanded to be provided with material safety data sheets and that a responsible chemical safety person be appointed. This exemplifies the monitoring of chemical transport in Finland. The Port of Kemi has created a rescue plan in case of an accident and rescue trainings have been carried out, even though the amount of hazardous substances being transported through the port does not require them. All environmental aberrations are reported to the port officers and in the case of large scale accidents, ELY is notified. All essential HSEQ-related information goes to the managers and the CEO, who make an overview and organizes trainings if needed.

The strategy and vision of the Port of Kemi are still a part of the city of Kemi strategy and they have not been updated recently. It might be worthwhile to review a possible update. Goals related to various indicators are documented in the quality manual required by the ISO 9001 (ISO 2008) certificate. These include goals related to HSEQ aspects as well as those related to administrative and economic indicators. The processes of the HSEQ issues are written down in the quality system and passed on to the employees. The companies operating in the port systematically share information regarding the HSEQ issues with new operators or employees. Mandatory meetings with Kemi Shipping are organized with an emphasis on health and safety and with carriage operators with a focus on quality and environmental issues. Quality is also discussed with the cleaning service provider.

The occupational instruction is carried out using a manual originally created for the city of Kemi and later modified to fit the port environment. Employees of the Port of Kemi Ltd. are interviewed concerning their job satisfaction and progress once a year. No systematic reward practices exist but casual rewards such as a free lunch might be in use. A customer satisfaction study is organized once every 2 years and has always provided initiatives and ideas for improvement. The focus used to be on marketing aspects but now it is more customized and includes the HSEQ issues if relevant to the customer. Feedback is also collected.

The facilities and working space are suitable and functional both for the clerical workers in the office and for the port officers. Construction and improvements are done on an ongoing basis as needed. Proper work equipment and systems are also provided, and they are considered functional with improvements made in various areas. The CEO gives permission and approves budgets for acquisitions by responsible employees. The orders are closely monitored.

3.2 Interviews of Representatives of Kemi Shipping

Kemi Shipping handles many tasks related to shipments moving through the port. All of the employees have the Finnish Occupational Safety Card (Väyrynen et al. 2008; Safety Card 2014) and internal and external training is offered continuously (e.g., for new equipment and machinery). Working hours are monitored with a resources management system and an overtime report is generated by the payment administration as part of occupational safety legislation. Staff trainings have been documented more closely during the last 3–4 years. Occupational healthcare service is bought from the private sector and the contract includes health check-ups and workplace inspections also required by law. Additional healthcare features prescribed by a doctor such as massage are financially supported, as a big part of the staff works in physically demanding working conditions. Health check-ups are mandatory once every 3 years for workers (other than clerical and managing employees) and workplace inspections are carried out more than once per year.

Kemi Shipping has two occupational safety representatives: one for stevedore workers (who are in the majority) and another for managers and clerical employees. An occupational safety manager is listed in the Finnish occupational safety personnel register as are all other members of the company's occupational health and safety committee. The goals, responsibilities and competencies of occupational health and safety personnel are written down in the action plan linked to the OHSAS 18001 certificate (BSI 2007).

Risk assessment is required by law and Kemi Shipping uses Riski Arvi, a web-based risk assessment program provided by the Ministry of Social Affairs and Health and the Centre for Occupational Safety (Basic Finnish tool 2013). About 30 jobs have been assessed and all new tasks go through the assessment process before implementation. Accidents are investigated to prevent their reoccurrence. Every accident is documented, as the information is needed for the insurance provider. Compulsory notification due to all severe accidents is submitted to the police and the Regional Authority of Occupational Protection. Notifications of dangerous situations by employees are reported and handled every week. Supplying companies' level of expertise in HSEQ is inspected as part of the Act on the Contractor's Obligations and Liability When Work is Contracted Out (TEM 2013). A big company with international certificates is assumed to have its HSEQ issues properly handled. Kemi Shipping appreciates a high level of expertise in HSEQ issues as a competitive advantage when deciding on business partnerships.

Kemi Shipping was granted the ISO 14001 certificate (ISO 2004) in 1999. Monitoring Finnish legislation and any possible changes in it has been outsourced to a consulting company. Kemi Shipping's management gets the information of new or changed legislation and acts accordingly. An environmental report has been done and all environmental deviation notifications are carefully studied on a weekly basis. Chemicals are listed and material safety data sheets are provided concerning all chemicals. Most chemicals are needed in the repair workshop, which produces most of the waste (such as oil waste) that is contracted to a specific waste handling company. The repair workshop has its own environmental permit while all other operations are under the Port's environmental permit (i.e. the permit of the Port of Kemi). When hazardous chemicals are transported or handled, all the other operators of the port and interest groups are notified. The environmental situation is monitored weekly via an observation walk-through by clerical employees and managers on the premises. This functions as an internal audit round with a varying subject of interest. Every 4 weeks, a special report is compiled to summarize the situation. The waste management company provides a monthly report on the screenings and the amounts are checked annually.

Quality is of major importance in Kemi Shipping, as shipping can be thought of as the final treatment of products leaving the Port. Quite much of Kemi Shipping's internal communication is about quality. It is seen as an aspect that can never be totally mastered, as there is always something that can be improved. Informing the staff about HSEQ issues is still seen as one thing that could be improved. Kemi Shipping has a management group whose meetings are documented. A managerial review is provided. The CEO has been creating a new strategy to replace the 5–7 year-old strategy. A summarizing folder for the occupational instruction and guidance was created 2 years ago. It provides a checklist of legal requirements and the issues that are important to Kemi Shipping. Employee satisfaction has been measured recently and is planned to be studied every 2–3 years. Rewards are offered by the initiative committee for making initiative suggestions and for reaching one month without any work-related accidents. Rewards can be in the form of a free lunch or a token for a cup of coffee and a bun at the Finnish Seamen's Mission in Kemi (which has been chosen as a target of support and goodwill by Kemi Shipping). Healthy activities outside of working hours such as going to the sawdust track, gym or bowling are financially supported and rewarded with the option of participating in a lottery of gift vouchers.

All the operating processes have been documented in the quality manual. Shipping requires specific equipment which means that about 70 different devices are used for specific purposes. The facilities are functional and clean. Cleaning has been outsourced and the level of tidiness is good; no complaints have been received related to the numerous shipments that are stored in Kemi Shipping's facilities before shipping. New technologies related to work do exist and implementing them is to be improved. A recent example is changing to a new operating system.

A resource planning information and communication (ICT) system handles orders and communicates with customer companies. HSEQ issues of the supplying companies are communicated daily but not necessarily documented. The idea to involve Havator to participate in the current research was proposed by Kemi Shipping, and should be seen as part of HSEQ-related communication. Developing and planning is seen as part of every staff member's job; however, managers and foremen in particular should observe how the level of safety and efficiency in the workplace can be continuously improved.

A customer satisfaction inquiry has been performed in cooperation with the Port of Kemi. Customer feedback is also gathered and possible reclamations are always processed. The daily operation of Kemi Shipping is measured using ten different indicators that have been documented in the management system and are monitored on an annual basis linked to the management's review. The indicators include HSEQ-related indicators such as work-related sick leave days, waste control, hazardous incidents and accident frequency.

3.3 Interviews of Representatives of Havator

During the interviews, it was noticed that Havator had already been evaluated using the HSEQ AP run by major process industry companies for some years ago; however, the assessment comprised all its functions in northern Finland, especially as far as a key supplier of the process industry companies, like steel mills. In this study, Havator is positioned as the last link of a supplier chain. It provides crane services in Finland, Sweden and Norway and the Baltic countries. Some project work is also done in Russia, usually in cooperation with another company that has had previous activity there. Demand in Russia is limited but is slowly growing and can be easily met by Havator, which has operating experience in Arctic conditions. The climate poses limitations to the use of machinery. The biggest challenges are those associated with complicated paperwork such as working permits and visas when transferring personnel to work in Russia.

In the Port of Kemi, Havator employs about five to six people. It is a relatively new operator in the Port but big investments were made to acquire the necessary machinery and cranes when it started in 2012. The personnel are qualified and all trainings are closely monitored and regularly repeated. As the working equipment is highly specific, internal trainings are provided constantly. Managers have separate trainings but all members of staff receive safety training as part of their work. Different work-related trainings such as those regarding the lifting work supervisor and safety harness inspector qualifications are also seen as necessary and are therefore provided. The other operators in the Port of Kemi require Havator's employees to hold the Finnish Occupational Safety Card (cf, Väyrynen et al. 2008; Safety Card 2014).

Occupational healthcare services have been purchased from one private healthcare provider; however, in southern parts of Finland separate contracts are also made. According to the law, health check-ups and workplace inspections are included in the contract but additional treatments such as physical therapy are partly sponsored by the company for the employees. The company has one occupational safety manager who supervises the all sites of Havator in entire Finland in close cooperation with the local managers and site managers who are better aware of the concrete facts and circumstances of each site. An occupational safety representative has been named for sites with over 10 employees and a "safety advocate" for sites with fewer employees.

Risk assessments are carried out for each site when changes occur and every 5 years even with no changes. The crane business requires that a lifting plan be made for each time lifting is done. All safety-related requests from the client are also noted and considered in the individual lifts. So far Havator has not encountered any major accidents and internal research has always been enough when accidents do occur. All investigations are documented and most of them have to do with the Nordic climate, e.g., snow and ice-related slips and falls. Hazardous incidents are also reported and presented in the meetings of the safety committee. In serious cases, information is spread and preventive actions are taken. Some of the cases are also reported to the employees in the journal published for the personnel. The ISO 9001 (ISO 2008) -related quality manual includes indicators related to the accidents and hazardous incidents which are published annually in the management review. Havator outsources very few of its operations and its subcontracting companies have been known for a long time and are checked to be competent enough in HSEQ issues.

Havator produces relatively little waste. Of its waste, oil waste is seen as the most problematic. Recycling and sorting of waste is not necessarily included in the orientation of new employees; however, all waste containers for different sections of waste are marked with a description of the waste as law requires. Sorting is monitored in the occupational health and safety rounds.

Documentation of toxic waste movement is kept for 3 years as required by law; other waste is monitored and documentation is available going back 5 years. The waste management provider offers waste statistics, although no goals are set for each section. The environmental effects of operating were studied accurately 6 years ago by a independent research company hired by Havator. Monitoring of legislation compliance related to Havator's operations has been outsourced. Updates are given every month and instantly when substantial changes take place. Environmental deviation notifications are gathered but are very rarely received (e.g. usually less than once a year). Updating the list of used chemicals and the material safety data sheets are the responsibility of the head of each site. The features of project work, such as a closed timeframe and changing circumstances, create challenges for monitoring the chemicals. Environment-related reporting is not substantial due to the low environmental harm caused by operating. Havator is more often the receiver of environment-related notifications than the provider of them. Project-related information is considered especially important.

The management board meets eleven times a year and a summary of HSEQ issues is included in the agenda. Internal communication is mostly focussed on basic occupational instruction but all employees are informed about the safety-related goals. The instruction has a base form but is seen as something that could be standardized, as the person in charge obviously has a great impact on the guidance presented during an instruction. Strategic goals such as budgetary goals and other goals are reported in monthly meetings.

Employee satisfaction is investigated once every 2 years. Development conversations with employees were implemented a couple of years ago and are still in progress. No agreed upon system of rewards has been adopted but the top management has a bonus system and workers are offered a free lunch for every hazardous near miss case they report. The facilities are functional and clean enough. As usual in the crane business, the equipment is specific and functional and back-up machinery exists. Keeping track of equipment such as spare parts is not a problem. Customer feedback is collected from each site and reviewed in monthly meetings. Feedback on customer satisfaction is performed after projects.

The processes have been described in the resource management system and working methods and technologies fulfil the company's needs. Plans and developments for operations have also been documented in the system and are influenced by the feedback and customer requests. Havator has created the quality manual as part of the ISO certification. The Turvallisuuskymppi system (Liuhamo and Santonen 2001), a basic Finnish safety management self-audit for suppliers, has also been implemented. HSEQ issues are seen as a constant target for development with partner companies. HSEQ indicators in the manual have been monitored more closely during the last 5–6 years. They are of special importance to foreign customers who want to evaluate the company's operations and competitive position. The need to monitor basic economic indicators is considered obvious for a quite big company like Havator. Every set of indicators is monitored weekly. Goals are set and revised if needed. An example of a metric that has reached its goal is the percentage of personnel absence, which has dropped below the target limit of 3 % of planned work time.

3.4 Summary of the Interviews of the Case Study Companies Operating in the Area of the Port of Kemi

The three companies interviewed gave an impression of having many of the HSEQ issues well managed and in control. That is largely due to the fact that all of them have been granted ISO 9001 (ISO 2008), ISO 14001 (ISO 2004) and OSHAS 18001 (BSI 2007) certification. In the conversations that took place during the interviews, all interviewees saw the certificates as an excellent prerequisite for potential passing the actual HSEQ AP. In addition, the relatively high demands of the Finnish legislation were seen as backing up many issues. Havator is the only company with real experience in the HSEQ AP.

All companies were in contact with others about the themes and information seemed to be shared on a daily basis, although not always officially or with documentation. The Port of Kemi seemed to be well aware of its role as the port keeper and its leading role in a shared workplace. Its role extended from occupational safety to environmental responsibility as well because the environmental permit had been granted. Havator seemed to be more in the role of information receiver. The forms of communication still seemed a bit blurry and were not clear in the interviews. It might be that some form of synergy is required. Cooperative studying of the current procedures and developing and clarification could be needed. This is something for which the HSEQ AP could be used.

In the manufacturing industry sector, a large scale production company, e.g. one of process industry, usually has smaller companies as suppliers. The size of each company is likely to decrease when going down the supply chain. In this case of the port, it should be noted that the situation is reversed: the Port of Kemi Ltd. is the smallest of the companies when measuring by turnover or employees. In 2012, the Port of Kemi had a turnover of 5.962 million Euros and about 30 employees, Kemi Shipping had a turnover of 17.157 million Euros and about 120 employees and Havator (the largest) had about a 110 million Euro turnover and about 600 employees.

To summarize, the three companies seemed to be quite up-to-date when it comes to health, safety, environmental and quality management in their operating fields. Company representatives took pride in talking about the areas they felt to be the strongest and when something felt a bit imperfect, corrective ideas and possibilities for improvements were covered with no shame or hiding. It was mentioned that the practice of sending the HSEQ AP questions to the assessed company beforehand (Väyrynen et al. 2012) was helpful for preparation for the assessment, not only to gather and organize the required proof and documentation but also to check the current status of things and create a small group of people who were best suited to answering the questions and most aware of any recent changes. This indicates that the opportunity to gain extra information from the actual HSEQ AP evaluators during the assessment would most likely be welcomed and the ideas and tips would be appreciated and put into use. It became clear during the interviews that the evaluators should be trained for performing the HSEQ assessment and additional knowledge and prior experience would be useful.

Providing a logistical link to the Barents area was described as a possibility but according to the companies it seemed to be highly unlikely to occur. The companies preferred working across the border mostly in partnership with companies that had already established business in Russian Federation. The biggest challenges were seen to be the excessive paperwork and the unpredictability of procedures related to working in Russia. The market seems to be there but there was no eagerness to reach it. All existing connections to Russia were project-related and there were not many (although their number is increasing). Even though port business is of course an international field of business as it operates on the border of a country, the most likely business partners were seen to be from other Scandinavian countries rather than from Russia.

3.5 Interview of a Representative of If Insurance Company

The interview of a specialist from the If insurance company was performed in January 2013. The point of the interview was to get an opinion about using this kind of evaluation method in the port industry from a stakeholder involved and fully acquainted with the port industry but not operating in port business. The insurance company finds measures taken towards improving occupational health, safety, environment and quality as very positive.

The ambiance of the interview was positive and encouraging. The outlook towards the HSEQ AP being adapted in the port industry was supportive and the benefits from implementing it were discussed in broad terms. The method could be seen as a "best available technique" because it is considered to bring added value to the field. Increasing cooperation and communication between four fields of HSEQ were discussed to be beneficial for all parties involved.

If's involvement in this study was welcomed. The interview and helpful attitude still showed belief in the future development of and research on the applicability of the HSEQ AP. Insurance companies can be expected to speak up for all innovations likely to reduce risks related to occupational health and safety as well as environmental problems or quality issues. That is why positive feedback towards systematic HSEQ improvements is seen as worth mentioning.

4 Discussion

Utilising the literature referred in the introduction on the issues of health, safety, environmental and quality management in general, the HSEQ AP used to assess those issues, the empirical findings describing the companies operating in the Port of Kemi, the following answers can be presented to the research questions posed at the beginning of this study.

1. How does the HSEQ AP work and what are its main principles?

The HSEQ AP encourages the company being assessed to look at its existing management systems of health, safety, environment and quality. After a self-assessment, the company can perceive what its strengths and weaknesses are in relation to the issues connected to the question set. The process of gathering proof and evidence of the current state of affairs makes the person or people in charge gather the latest relevant information and become aware of all HSEQ aspects. At this point, the company can still make small changes and corrections, if necessary (Väyrynen et al. 2012).

In the actual assessment, the company is assessed and provided with an outside opinion concerning the current status of its HSEQ. Getting an outsider's opinion ensures that the evaluation will be neutral and in no way biased (as internal audits occasionally may be). Even with documentation gathered and presented to the evaluators, the company's own perception and the assessment team opinions are likely to differ. The company's representatives are offered some guidance and tips on sections that are found to be defective. The evaluation is not developed and performed simply to show the assessed companies their shortcomings and weaknesses, but rather to help them improve the necessary areas or to offer different ways of looking at things. This means that the changes made are always a result of both self-assessment and external assessment. After the actual assessment, the HSEQ AP provides support for any possible changes with a reassessment and follow-up. The HSEQ AP eliminates the need for separate external health, safety, environmental and quality audits performed by multiple customer companies that are likely to include same indicators and questions (Väyrynen et al. 2012).

2. Can the HSEQ AP be used to create solutions with additional value to the port industry?

Principally, the HSEQ AP could provide the same benefits to the port industry as it does to other industrial sectors: it reduces the amount of external audits. The HSEQ AP serves not only as an external audit but also an internal audit in the form of the self-assessment. Expertise in the HSEQ areas would likely be passed down the supplier chain and good practices and knowledge would be shared in the assessments. This would lead to a rising amount of companies that are successful in all HSEQ aspects. A port operates with many different kinds of companies. The HSEQ AP would make it easier to compare supply companies and if agreed upon, ports could benchmark each other in order to foster collaborative learning.

The high level of HSEQ issues should also be seen as an asset in market competition. It can be seen as an advantage in competition between ports and between sea transport and other means of transport. The HSEQ AP would be a good stimulus for competition that is not likely to have any negative impacts on the company's operation and would encourage continuous improvement.

3. What kind of changes would the HSEQ AP bring to the Port of Kemi?

Three participating companies of this study operating in the Port of Kemi have all paid attention to all main aspects of HSEQ. This opinion is based on the existing relevant standards they all follow and information and impressions gathered in the interviews. Integrating the HSEQ Assessment Procedure into management systems of these three companies would further improve their level of performance and expertise and increase the amount of cooperation and communication between companies. It might attract more partners who have been using other methods or transport suppliers. If the Port of Kemi, Kemi Shipping, and Havator would all apply themselves, and promote and encourage their suppliers to consider and implement the HSEQ Assessment Procedure, the expertise would spread even further. The Assessment Procedure would be likely to decrease the number of external audits, especially if more than one of the existing companies implemented the Assessment Procedure. The HSEQ AP could also be promoted as an indicator of interest in the HSEQ field and possibly even as an achievement similar to the ISO certificates.

4. What can be said about the applicability of the HSEQ AP to the port industry, according to this study?

The HSEQ AP received good feedback in the interviews from all the participating companies in this study. The HSEQ AP was found thorough but not too accurate. The companies saw the Assessment Procedure as very positive and as a good fit for the port industry but of course some alterations might be necessary. As Havator has been HSEQ assessed previously, it is a good example of a company that is not exclusively operating in ports but still acting as a critical player in the Port of Kemi. This indicates that industry and ports might share some suppliers and thus ports might already have suppliers with the HSEQ Assessment Procedure in use. The current trend of incorporating ports is a great opportunity for ports to update and modernize their practices. The HSEQ Assessment Procedure would offer a good endorsement for contemporary overall status.

The port industry has a strong foothold in the Finnish economy as Finland has a history of export. As mentioned in the interviews, the port can be seen as an extension of the production line of a customer company as goods are handled, possibly stored and packed for transport. Transport should not be seen as a necessity of doing business but as an asset in competition as it offers possibilities for improvement of deliveries. Poor quality, lack of occupational health and safety processes, or environmental practices are noticed not only by the company itself but also by its customers. Investing in HSEQ areas can be beneficial to the competitiveness of the company, the work atmosphere, well-being at work and the incidence of sick leaves or accidents. Thus, it would most probably reduce costs related to decreased working capacity, occupational health or accidents and getting substitutes for personnel on leave. This makes it important for the customers of the port to get proof of good skills and expertise in structured handling of health, safety, environmental and quality management.

The likelihood of increased business in the Barents area is great if the current global warming frees the Northern Sea Route for a more active use. It is obvious that with more traffic and activity, the Arctic will be exposed to greater environmental risks. A high level of environmental management would help to reduce the risks and soften the impact of heavy transport and industry. By removing some of the obstacles and easing the cross-border traffic by land between Russia and its north-western neighbours, the load of sea traffic maybe be reduced.

This experience of using the HSEQ AP in the port industry proves that the HSEQ AP can be used outside the process industry. It could be beneficial across many different types of companies but will need to be adjusted accordingly. Based on the research results, it can be stated that the HSEQ AP can be adapted to the port industry; however, it requires further investigation using multiple port cases. Repeating the interviews or performing actual assessments might provide useful information on the fits and misfits in the AP. As for the interviews in this study, the expertise of the interviewer was not as high as of those performing the actual assessments and some misinterpretations or misunderstandings might have altered the results. It must also be noted that Finnish ports can be quite different from each other, as some ports control almost all port operations and services and some ports have outsourced their operations.

The ongoing trend of public utilities becoming incorporated companies might be a good area to study in regards to the companies' existing processes. The procedure of a communal company becoming incorporated is likely to be good timing for an audit of the existing activities, as changes of management are likely to take place in all levels of operations. Incorporation might also leave companies without new ideas to replace the practices used in the previous business form that felt in need of change.

Companies in Finland are required to pay a lot of attention to the fields of occupational health, safety, environment and quality management. For example, laws already exist that require insurance to cover all employees and occupational health care to be provided for everyone. Due to the high level of expertise, an assessment method created in Finland has the potential to be published internationally and circulated between the neighbouring countries. Currently some of the requirements mentioned in ISO standards are already covered by the demands of the legislation and are self-evident.

Safety is a multifaceted concept. During this research, the environmental issues linked to safety were of key importance and should not be forgotten when discussing safety. Including security as a part of safety or raising it even equal to it was already of interest when this study began. In ports, physical security has already been included in the ISPS codex, which includes fencing the port area and close monitoring of traffic in and out of the port, both on land and by sea. The port has not encountered any outrage or troublemaking and none of the companies described any extra need for physical security. As the Port of Kemi is not located right next to the city, it might be safe from public nuisance. Security includes not only the gate keeping and fencing but possibly a video monitoring system, decent lighting and locking systems. As no problems in this area have occurred, further study is not necessary at this time (but is still recommended for some point in the future).

Nowadays, virtual security must be seen as a critical part of security. There has been an increase in electronic information systems being broken into. This should be seen as troublemaking or a physical offense. In some cases, companies rely on the common sense of employees who are trusted not to post confidential information or to pass on any business secrets. Putting virtual security codes and instructions into operation was not mentioned in the interviews but some guidance had been given to the new employees. In our time, online activity has increased dramatically, which might eventually blur the lines between what is accepted and not accepted. Therefore, this might still be a good area to investigate even though it is not considered necessary at the moment. Guiding leaflets or online security advice might be included in the introduction and reminders are given every once in a while.

Management of virtual security related to electronic information and ICT systems might also be a potential aspect to be added to the HSEQ AP in the future. Its challenges might include that companies of different sizes tend to have significant variation in their levels of use when it comes to online programs and ICT services. This option for developing HSEQ AP should be investigated at some point. Virtual hazards should still be considered as everyday risks when conducting business.

Getting the opportunity to visit the Commercial Port of Murmansk and trying to compare it directly to the Finnish ports was one of the remarkable experiences of this study. Unfortunately, the language barrier and a tight schedule were not conducive to focussing on the separate fields of HSEQ in general, or HSEQ AP specifically, in the Commercial Port of Murmansk. Nevertheless, it was a clear impression that, for example, the local enterprise and safety culture contains a lot of differences compared to Finland. According to the World Bank (2014), two indices describing the level of port infrastructure and facilities were given for three countries of the area: Finland, the Russian Federation and Sweden. The indices were as follows:

- Logistics performance index in 2012, scores 1–5 (worst … best): 4.05; 2.58; and 3.85 for Finland, the Russian Federation and Sweden, respectively.
- Quality of port infrastructure in 2013, scores 1–7 (worst … best): 6.4; 3.9; and 5.8 for Finland, the Russian Federation and Sweden, respectively.

Perhaps Finnish companies could be used to set an example for Murmansk regarding how these areas are managed. Maybe this would also spark an interest in knowledge exchange, which is one of the goals of the Barents Logistics 2 project. At the national level, wide utilisation of the HSEQ-related auditing or assessing systems like HSEQ AP could raise the general level of the ports' management, facilities and services in a way that could contribute to improved performance and perceptions in national level surveys like the one from the World Bank (2014).

The Port of Kemi is actually, like many Finnish ports, quite closely and predominantly tied with the process industry supply network. That is one reason for supporting the introduction and implementation of HSEQ AP in the Port of Kemi. Another supporting view might be the one of the entire maritime sector with shipyards, transport by water (shipping companies) and harbours: the whole cluster could be one basis for joint developments in the field. For example, Singapore shipbuilding and ship-repairing is given a very thorough HSE instruction (Workplace Safety 2009). The Norwegian shipping industry's safety management performance has been analysed for accelerating developments concerning tankers (Oltedal 2010). The offshore oil industry might be linked with this possible huge cluster of synergies. The offshore oil industry has a long tradition of enhancing and guaranteeing high performance in HSEQ management (Wagenaar et al. 1994; Kjellen 2000).

On the other hand, it seems that ports are not always so frequently included in the maritime context. While going through the programme of the Applied Human Factors and Ergonomics (AHFE) 2014 Conference (which had more than 1,400 participants) (Ahram et al. 2014), one could found 26 maritime-related papers, mainly in the sub-conference "International Conference on Human Factors in Transportation." Furthermore, none of those papers were related to work activities in ports. Typical topics consisted of issues like onboard navigational systems, offshore simulator work training and the ergonomic user-interface design of ships. Probably the most (though indirectly) port-related paper dealt with the handling of shore-based unmanned ships.

This study comprises an exercise in how HSEQ AP brings good properly scaled practices of economic, social and environmental sustainability to small and medium-sized companies of the seaport industry (cf, Zink and Fischer 2013; Väyrynen et al. 2014; Worldsteel 2013).

This study uses the concept of HSEQ. On the other hand, according to Harms-Ringdahl (2013), the expression Safety, Health, Environment, Production (SHEP) or HSEP might be equally in use. This study emphasised to promote adding positive issues to production while simultaneously preventing undesired consequences. So Harms-Ringdahl's (2013) definition of P ("to prevent problems with production, quality, etc., and loss of property") is a good one as well. Qualitative descriptions of production are not enough. Figures and facts are needed as well for follow-up, benchmarking and managing in companies. HSEQ AP is able to provide the maritime sector with quantitative indicators that are needed for evidence of guaranteeing sustainable progress (Väyrynen et al. 2012, 2014; Zink and Fischer 2013).

Multiple management systems (MMS) are closely related to integrated management systems and standards, and somewhat related to quality award models (EFQM 2013). The current MMS standard (ISO 2011) describes how to "survive" better, because in today's business environment, many organizations incorporate a number of management systems such as quality, environmental, ICT services and information security. These organizations want to harmonize and, where possible, combine the auditing of these systems. Leading business continuity experts wrote a new standard entitled "Societal security–Business continuity management systems–Requirements." This standard provides the best framework for managing business continuity in an organization, which can become certified by an accredited certification body (ISO 2012). Its main approach is that business continuity is part of overall risk management in a company, with areas that overlap with information security management and IT management.

Regarding future research for the northern seaport industry, maybe it should be studied whether good criteria could be built by collecting the best features of the following approaches: ESPO (2012), Brisbane (Health and Safety 2013), and HSEQ AP. These specific features could provide seaport industry with a sound basis to be enhanced by and combined with the coming general international standardised guidelines for occupational health and safety management (ISO 2014).

5 General Conclusion

Seaports can be characterized as specific and quite challenging places to work. They provide key services. Supply chain competitiveness and effective risk control actions within whole logistics and supply chain and partner networks, including well-being at work, are the goals of holistic management in seaports. Well-being and productivity at work, i.e. high enough quality of working life, is a more and more crucial factor in all competitive industrial activities. As far as risks, the well-known ALARP (As Low As Reasonably Practicable) approach has to be applied and realized in practice as a process of continuous improvement. As far as societies and citizens are concerned, it is always a question of continuity management by reliable sea logistics. Many companies with their valuable tangible assets are customers of seaport services. Confidence is needed to provide for governmental and private organisations.

All the above, H(ealth), S(afety), E(nvironment) and Q(uality) belong to the field of contemporary holistic quality control, assurance and management measures. Integration of all aspects of quality is a practical, rational, feasible and cost-effective model.

HSEQ management is for fulfilling both legal and business requirements. Developments towards balanced sustainability are simultaneously promoted. Key roles and requirements belong to employers' responsibilities for ensuring that they are adequately managing health, safety and environment in the context of holistic quality management. The Finnish process and manufacturing industry is utilizing

HSEQ AP to encourage prerequisite actions for guaranteeing success in its value networks. Both core and assisting partners are contributing and collaborating in modern seaports. The key operator starts cooperation and coordination with other employers sharing a workplace. Sharing responsibilities of the many employers and contractors involved in this industry is an essential and forced feature of this service industry. This means that communication of combining business and HSEQ issues is critical.

References

Borgerson SG (2008) Arctic meltdown: the economic and security implications of global warming. Foreign Aff 87(2):63–77

BSI (The British Standards Institution) (2007) BS OHSAS 18001 occupational health and safety management. BSI, London

EFQM (2013) EFQM excellence model. http://www.efqm.org/the-efqm-excellence-model. Accessed 08 Apr 2014

ESPO (European Sea Ports Organisation) (2012) Towards excellence in port environmental management and sustainability. http://www.espo.be/images/stories/Publications/codes_of_practice/espo_green%20guide_october%202012_final.pdf. Accessed 08 Apr 2014

European Agency for Safety and Health at Work (2011) OSH in figures: occupational safety and health in the transport sector—an overview. Publications Office of the European Union, Luxembourg. doi:10.2802/2218

Harms-Ringdahl L (2013) Guide to safety analysis for accident prevention. IRS Riskhantering AB, Stockholm

Havator (2014) http://www.havator.com. Accessed 08 Apr 2014

Health and Safety Port of Brisbane (2013) http://www.portbris.com.au/about-us/health-and-safety. Accessed 08 Apr 2014

Hirsjärvi S, Remes P, Sajavaara P (2007) Tutki ja kirjoita (Study and write), 13th edn. Tammi, Helsinki

Hugnes P, Ferrett E (2003) Introduction to health and safety at work. Elsevier Butterworth-Heineman, Oxford

International Organization for Standardization (ISO) (2004) ISO 14001: environmental management systems—requirements with guidance for use. ISO, Geneva

International Organization for Standardization (ISO) (2008) ISO 9001: quality management systems—Requirements. ISO, Geneva

International Organization for Standardization (ISO) (2010) ISO 26000: guidance on social responsibility. ISO, Geneva

International Organization for Standardization (ISO) (2011) ISO 19011: guidelines for auditing management systems. ISO, Geneva

International Organization for Standardization (ISO) (2012) ISO 22301: societal security—business continuity management systems—requirements. ISO, Geneva

International Organization for Standardization (ISO) (2014) ISO/CD 45001: occupational health and safety management systems—requirements. ISO, Geneva

ISRS (International Sustainability Rating System) (2014) ISRS—for the health of your business. http://www.dnvba.com/Global/sustainability/management-practices/Pages/isrs.aspx. Accessed 16 May 2014

Karvonen T, Tikkala H (2004) Satamatoimintojen kehittäminen ja satamia koskevan lainsäädännön uudistaminen (Development of harbour activities and legislation). Publications of the (Finnish) Ministry of Transport and Communications 65/2005

Kemi Shipping (2014) http://www.kemishipping.fi/?lang=en. Accessed 08 Apr 2014

Kjellen U (2000) Prevention of accidents through experience feedback. Taylor & Francis, New York

Liuhamo M, Santonen M (2001) Turvallisuuskymppi (SafetyTen). The Centre for Occupational Safety, Helsinki

Lun YHV, Lai K, Cheng LTE (2010) Shipping and logistics management. Springer, London

Managing (2002) Managing health and safety in dockwork. http://www.hse.gov.uk/pubns/books/hsg177.htm. Accessed 08 Apr 2014

Ministry of Social Affairs and Health (2013) Occupational safety and health in Finland. http://www.stm.fi/c/document_library/get_file?folderId=1087418&name=DLFE-13602.pdf. Accessed 08 Apr 2014

Oltedal HA (2010) The use of safety management systems within the Norwegian tanker industry—do they really improve safety. In: Bris R, Guedes Soares C, Martorell S (eds) Reliability, risk and safety: theory and applications. Taylor & Francis Group, London, pp 2355–2362

Port of Kemi (2014) http://www.keminsatama.fi/en/home/. Accessed 08 Apr 2014

Qvist S (2010) Kalustonhallintajärjestelmän käyttöönotto ja käytön kehittäminen (Implementation and development of facility and machinery management system). Kemi-Tornio University of Applied Sciences, Department of Industrial Management, Kemi

Riski Arvi (2013) Basic Finnish tool for risk assessment. http://www.ttk.fi/files/2941/Riskien_arviointi_tyopaikalla_tyokirja_26022013_TTK.pdf. Accessed 08 Jan 2013

Safety Card (2014) http://www.tyoturvallisuuskortti.fi/files/206/Occupational_Safety_Card_TTK.pdf. Accessed 08 Apr 2014

Safety in Docks (2014) Safety in docks: approved code of practice and guidance (ACOP), L148. http://www.hse.gov.uk/pubns/books/l148.htm. Accessed 08 Apr 2014

Sinclair MA (2005) Participative assessment. In: Wilson J, Corlett N (eds) Evaluation of human work, 3rd edn. CEC Press, Taylor & Francis Group, London, pp 83–111

TEM (2013) Contractor's obligations and liability. http://www.tem.fi/en/work/labour_legislation/contractor_s_obligations_and_liability. Accessed 08 Apr 2014

The Barents Euro-Arctic Council (2013) Barents Euro-Arctic Region. http://www.beac.st/in_English/Barents_Euro-Arctic_Council/Partners_and_related_organisations.iw3. Accessed 14 Jan 2013

The Barents Euro-Arctic Region (1993) The Kirkenes declaration. In: Paper presented at the conference of foreign ministers on cooperation in the Barents Euro-Arctic Region, location of conference, 11 Jan 1993

The Finnish law on waste controlling (646/2011)

Väyrynen S et al (2008) Finnish occupational safety card system: special training intervention and its preliminary effects. Int J Tech Human Interaction 4(1):15–34

Väyrynen S, Koivupalo M, Latva-Ranta J (2012) A 15-year development path of actions towards an integrated management system: description, evaluation and safety effects within the process industry network in Finland. Int J Strategy Engr Asset Mgmt 1(1):3–32

Väyrynen S, Jounila H, Latva-Ranta J (2014) HSEQ assessment procedure for supplying industrial network: a tool for implementing sustainability and responsible work systems into SMEs. In: Ahram T, Karwowski W, Marek T (eds) Advances in human factors and ergonomics 2014. Proceedings of the 5th AHFE conference, Krakow, Poland, 19–23 July 2014, pp 6570–6580

Verny J, Grigentin C (2009) Container shipping in the Northern Sea route. Int J Prod Econ 122(1):107–111

Wagenaar WA et al (1994) Promoting safety in the oil industry. Ergonomics 37:1999–2013

Workplace Safety (2009) Workplace safety & health manual for marine industries. https://www.wshc.sg/wps/themes/html/upload/cms/file/WSH_Manual_Marine_Industries.pdf. Accessed 08 Apr 2014

World Bank (2014) World development indicators: trade facilitation. http://wdi.worldbank.org/table/6.7. Accessed 16 May 2014

Worldsteel Association (2013) Worldsteel Association—sustainability. http://www.worldsteel.org/steel-by-topic/sustainable-steel.html. Accessed 08 Apr 2014

Yin RK (1989) Case study research: designs and methods, 4th edn. Sage, Newbury Park

Zimmerbauer K (2013) Unusual regionalism in Northern Europe: the Barents Region in the making. Reg Stud 47(1):89–103

Zink KJ, Fischer K (2013) Do we need sustainability as a new approach in human factors and ergonomics? Ergonomics 56(3):348–356

Chapter 7
Truck Drivers' Work Systems in Environments Other Than the Cab—A Macro Ergonomics Development Approach

Arto Reiman, Seppo Väyrynen and Ari Putkonen

Abstract Local and short haul (L/SH) drivers work in various other work environments in addition to that of the truck cab, and the safety of these environments vary widely. In this study, we combined methods from three time perspectives; accident statistics analyses (past); video observations (present) and scenario workshops (future) in order to provide new knowledge that can be applied to design and management process development in the transportation industry. Even though new technologies have and will emerge to ease drivers' work, the work that is performed in environments other than truck cabs still involves tasks that require physical activities and pose risks of occupational diseases and accidents. Thus, drivers' safety at work and work ability issues remain an area that needs continuous, systemic development. The results inevitably show that in order to successfully improve L/SH drivers' work, the relevant stakeholders' participation and a systemic approach is crucial.

Keywords Accidents at work · Local and short haul · Macroergonomics · Participative ergonomics · Safety at work · Truck driver

---

A. Reiman (✉)
Finnish Institute of Occupational Health, Aapistie 1, 90220 Oulu, Finland
e-mail: arto.reiman@ttl.fi

S. Väyrynen
Faculty of Technology, Industrial Engineering and Management, University of Oulu,
P.O. Box 4610, 90014 Oulu, Finland
e-mail: Seppo.Vayrynen@oulu.fi

A. Putkonen
Turku University of Applied Sciences (TUAS), Sepänkatu 1, 20700 Turku, Finland
e-mail: ari.putkonen@turkuamk.fi

© Springer International Publishing Switzerland 2015
S. Väyrynen et al. (eds.), *Integrated Occupational Safety and Health Management*,
DOI 10.1007/978-3-319-13180-1_7

1 Introduction

Local and short haul (L/SH) truck drivers work in many different work environments in addition to that of the truck (see Fig. 1). As well as driving work, tasks such as loading and unloading cargo are performed frequently during the work shift (Olson et al. 2009). Different studies estimate that roughly about one-third of the work in L/SH operations involves driving; the rest of the work shift is spent on other assignments, carried out somewhere other than in the truck cab (see e.g. Hanowski et al. 2000). L/SH operations are thus a combination of both static work postures while driving and manual work assignments, such as loading and unloading cargo (Wioland 2013). A high prevalence of work-related musculoskeletal disorders and high accident rates have both been widely associated with the L/SH industry (van der Beek 2012; European Agency for Safety and Health at Work 2010; Miilunpalo and Olkkonen 2013; Shibuya et al. 2010; Smith and Williams 2014). According to several studies (see e.g. Shibuya et al. 2010; Wioland 2013), the vast majority of occupational accidents occur while performing work outside the cab.

Work environments in which L/SH work is performed outside the cab include, for example, the transportation companies' own terminals or distribution centres, customers' premises and yards, public streets and places close to the truck such as cargo spaces, trailers and tailgate loaders (Olson et al. 2009; Shibuya et al. 2010). The work that is performed in these environments is manual. Sometimes technical equipment is used to ease the workload (Keyserling et al. 1999; McClay 2008). Usually drivers work alone in these work environments, and their safety is not fully

Fig. 1 L/SH work outside the cab is performed manually in several different environments; in this case in the cargo space, on the tailgate loader on a public street, and finally inside the customer's premises

monitored. In cases of accidents, help is not always quickly available (European Agency for Safety and Health at Work 2010). L/SH operations are strongly dependent on the health and wellbeing of the drivers. Huang et al. (2013), for example, suggest that the way in which a company's safety climate influences such lone workers should be studied further.

Additional regular risk factors in Nordic countries that affect employees' performance are changing environmental conditions, especially in wintertime. Drivers have to continuously move back and forth during the work shift from warm truck cabs to cold outdoor work. Changing environmental conditions and activity levels require continuous thermoregulatory adjustments and cause stress for the employee (Risikko 2009). Such climatic factors might also be associated with musculoskeletal symptoms, as pointed out by van der Beek (2012).

The above-mentioned combination of individual drivers, work tasks, work environments and technologies can be seen as a work system. A work system comprises a combination of people and technological equipment within a space and environment, and the interaction of these components within a managed organization. Some work systems are engaged in transactions with other work systems, and thus, constitute complex work system entities (Hendrick and Kleiner 2001; Kleiner 2008; Smith and Carayon-Sainfort 1989). Certain elements of the work system cause physical, psychosocial and cognitive loading for humans (Carayon 2009). Humans' physiological and psychological reactions to these loads often have a detrimental effect, which can reduce individual performance and lead to a greater propensity for human errors and violations (Kraemer et al. 2009).

Ergonomics aims to optimize work systems, as regards performance and effectiveness, without detriment to health, safety or other wellbeing factors at work. Different authors (see e.g. Dul and Neumann 2009) have criticized ergonomics as being typically managed by occupational safety and health (OSH) professionals at the company level. Very often, the link between OSH and company strategy is missing. However, according to an international standard of work systems (EN ISO 6385 2004), work system optimization may be evaluated on the basis of three categories (1) health and wellbeing, (2) safety and (3) performance (the quantity and quality of production). According to this holistic thinking, occupational risks threaten both factors of wellbeing and productivity at work (EN ISO 6385 2004; Dul and Neumann 2009).

Optimal work systems can promote wellbeing and productivity. Microergonomics is a framework that focuses on distinct components in a certain workplace (Hendrick 2007). Macroergonomics in turn is a systemic and holistic framework for understanding and improving complex work systems in which different stakeholders act together (Hendrick 2002). Ergonomics can be applied to companies' management processes in several different ways and by different stakeholders, if their benefits are understood. A systemic, macroergonomics view provides one framework for conceptualizing the complexity of work systems and their interactions. As in modern safety culture research (see e.g. Reiman and Rollenhagen 2014) human, technological, organizational and cultural factors and their interactions must also be understood in practical ergonomics development activities in order to make them effective.

Participatory ergonomics maybe considered one of the primary approaches to macroergonomics (Brown 2002). In participatory ergonomics, relevant stakeholders are invited to participate in problem solving (Kuorinka 1997; Vink et al. 2008). The European Agency for Safety and Health at Work (2012), Gyi et al. (2013), and García Acosta and Lange Morales (2008) provide examples of participatory ergonomics interventions in the field of logistics. Often macroergonomics problem-solving processes also automatically help solve practical microergonomics issues at work (Kleiner 2006). Safety is a key factor in ergonomics approaches, and, as pointed out by Morel et al. (2009), must be strongly emphasized in both micro- and macroergonomics development processes in order to gain holistic results.

This study discusses how participatory ergonomics was utilized in a Finnish project in which different stakeholders systemically analysed the risks and ergonomic discomforts in L/SH operations. The project contained different micro- and macroergonomic sub-processes. The main objectives of this study can be presented as follows:

1. To provide a short review of accident statistics in the Finnish transport industry.
2. To present the findings of the participatory ergonomics approach.
3. To discuss the future scenarios of L/SH drivers' work in 2020.

2 Study Design and Methods

Both observational and archival studies are needed in order to understand the full complexity of a natural environment, also when the possibilities for controlled experiments are limited (Bisantz and Drury 2005). The study employed methodological triangulation of accident statistics, video observations and scenario workshops, covering all three time perspectives of the topic; past (accident archival), present (observations) and future (scenarios), as presented in Fig. 2.

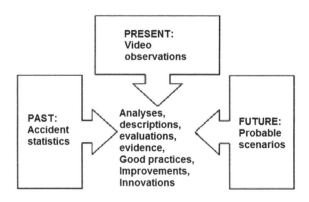

Fig. 2 The three dimensions of the development process

This study combines the frameworks of design science and ergonomics. Design science is a technology-orientated discipline that seeks to improve the human condition (Hevner and Chatterjee 2010; Järvinen 2007). In contrast, ergonomics is—in its premises—a human-centred discipline that focusses on improving work system design and management (Karwowski 2005). Three different methods and time perspectives, past, present and the future, (see Fig. 2) were utilized in order to provide holistic design knowledge for improving work systems in environments other than the truck cab in L/SH operations.

2.1 Accident Statistics Analyses (Past)

National accident statistics and verbal accident descriptions were retrieved from the Finnish Federation of Accident Insurance Institution's (FAII) database. The FAII database allows restricted access to researchers. An open coding approach (Strauss and Corbin 1998) was used to summarize and synthesize the accident statistics and descriptions data. The analysis covered all accidents that occurred to road transportation sector employees somewhere other than in truck cabs, and which were reported to insurance companies in Finland in 2006.

The accidents were classified into five categories: I. Ascending into or descending from the truck (including cab, cargo space, tailgate loader, outer body structures); II. Manual materials handling; III. Fastening and unfastening the load, IV. Manual installations and refuelling and V. Maintenance. These categories were based on the physical work activity that had led to the particular accident. Out of a total of 3,507 accident descriptions, 2,880 contained adequate information for this study's purpose.

2.2 Participatory Observations (Present)

Participatory observations of ergonomic discomforts and accident risks at work were carried out using the video-based observation method VIDAR (Video- och datorbaserad arbetsanalys in Swedish) (Kadefors and Forsman 2000; Forsman 2008). This method is mainly used for assessing general workload (Takala et al. 2010), and is based on previously recorded video material of an employee performing daily (routine) work. In the analysis sessions, the video material, which can be hours long, is condensed into a limited number of ergonomic problems at work (Kadefors and Forsman 2000). As the analysis is performed by humans, the method is subjective by nature (Forsman 2008).

The researcher followed and filmed the drivers' work in environments other than the truck cab during their L/SH operations in late winter 2008. Nineteen professional short-haul drivers [17 male, two female; average age = 31.5, standard deviation (SD) = 9.0 years; average driving years = 8.6 years, SD = 8.0 years] volunteered to take part in the individual analyses and eight male stakeholders (three immediate superiors, three industrial safety group members and two designers from a cargo space manufacturing company) took part in the group analyses. About 1,500 min of video footage of drivers' work was recorded and analysed.

The analysis identified three demanding work situations (see example in Fig. 3); (1) physical discomfort or (2) psychosocial discomfort, including hazard risks or (3) a combination of physical and psychosocial discomfort. A report of the recorded situations could be printed directly after the analysis. This report included a picture frame of a situation, observations and the frequency of the occurrence of the situation during a certain period.

In this case, the evaluator (i.e. the driver) identified a physical discomfort that was related to climbing on to the tailgate loader. Icy and snowy ground was identified as an additional risk to physical discomfort. According to the driver, similar work situations occurred four times per hour during the work shift. On the right of the figure is an additional scale that can be used for assessing the degree of discomfort of different body parts. This feature was not used in this study.

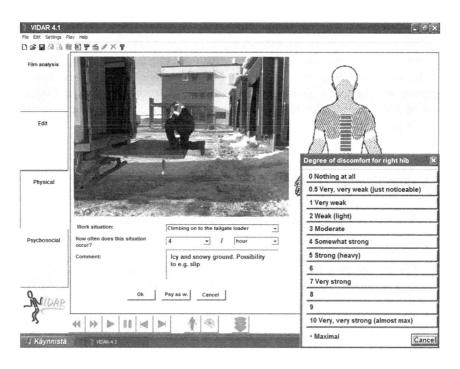

Fig. 3 The analysis frame in VIDAR of identified physical discomforts

2.3 Scenarios (Future)

A scenario workshop was arranged for the relevant stakeholders of the road freight transportation sector: the National Employers' Federation of Road Transport, the Transport Workers' National Union, the Organization of Transport and Logistics, the Technology Industry, the Ministry of Social Affairs and Health, the Insurance Sector, the Finnish Institute of Occupational Health, two national universities, two companies providing L/SH operations, two companies from the mechanical and ICT sectors, and funding organizations.

The practical microergonomics issues, in the form of the results of the video observations and accident statistics analyses, were introduced to the participants before the workshop. The workshop constituted a mixture of different techniques; brainstorming and 635 methods (cf. Pahl and Beitz 1988) with multiple phases including individual-, pair-, and group-based oral and literal approaches. More generally the approach included aspects of a five-step method (Ulrich and Eppinger 2004), including an external and internal search; the former being linked to participation by stakeholders representing, for example, various transport cluster partners (Lindhult 2008). In the workshop, participants ($n = 14$) formed future scenarios of delivery transportations in Finland in 2020. The scenarios dealt with improvements and new ideas related to technology, work systems, social issues, businesses, and services.

3 Results

3.1 Accident Statistics in the Transportation Industry (Past)

Figure 4 presents the five categories of accidents ($n = 2,880$). The vast majority of the accidents in the transport industry were related to manual material handling ($n = 1,630$, 57 %), followed by accidents related to physical activities in manual installations and refuelling. Manual installations consist of work tasks such as attaching trailers to the truck, installing cargo space supporters and installing walls to separate the cargo space.

3.2 Participatory Ergonomics Observations (Present)

The video observation sessions produced a total of 262 identifications of discomfort, i.e. demanding work situations. The majority of these ($n = 197$, 75 % of all identifications) were found in the individual analysis sessions by the drivers. Nearly half ($n = 97$, 47 %) of the identifications made by the drivers were related to demanding work situations that were performed in customers' yards and premises

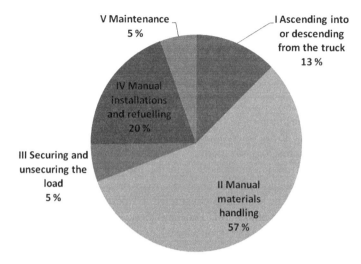

Fig. 4 Accidents ($n = 2{,}880$) in the Finnish road transport sector in 2006 classified by the physical specific activity to which they were related

or public places. In the stakeholder group analyses, 76 % of the demanding situations were related to tasks performed in cargo spaces, on tailgate loaders, in cabs and in other outer body structures (see Table 1).

The majority of the identifications concerned physical discomfort (149 identifications; 57 %) in different work situations, which occurred when performing tasks such as moving in and out of the cargo space and during manual handling of materials. Examples of psychosocial discomfort (113 identifications; 43 %) were related to factors such as time pressure, risks or obstacles in customers' premises, incorrectly functioning tools, and fear of causing damage or loss (to oneself, other people or material) as a result of one's own choices and actions.

The results of the video analyses show, interestingly, that the majority of the different stakeholders' identified discomforts were related to work performed in trucks (77 %); especially inside the cargo space or cab, on tailgate loaders or on the

Table 1 Identifications of demanding work situations ($n = 262$) by work environment in which they were performed, individual analyses and group analyses

	Customers' yards and premises and public places (n)	Trucks (including cargo space, tailgate loaders, cabs and outer body structures (n))	Home base (n)	Total (n)
Drivers' analyses	92 (47 %)	75 (38 %)	30 (15 %)	197 (100 %)
Stakeholders' analyses	5 (8 %)	50 (77 %)	10 (15 %)	65 (100 %)
In total	97 (37 %)	125 (48 %)	40 (15 %)	262 (100 %)

outer body structures of the truck. However, in the drivers' analyses, only 38 % of the identified discomforts were related to such work environments. These results should be reviewed rather cautiously, as the subject population in the stakeholder group is rather small and includes cargo space designers.

3.3 Scenarios Concerning L/SH Truck Drivers' Work in 2020 (Future)

The accident statistics analyses and participatory ergonomics observation results served as background microergonomics information for the future scenarios workshop that was arranged for relevant stakeholders. They worked together in order to obtain a systemic, holistic picture in order to understand and improve the complex work system in question.

The workshop results were divided into four main categories: strategic, national and international trends; companies; technologies and tools; and drivers, as presented in Fig. 5.

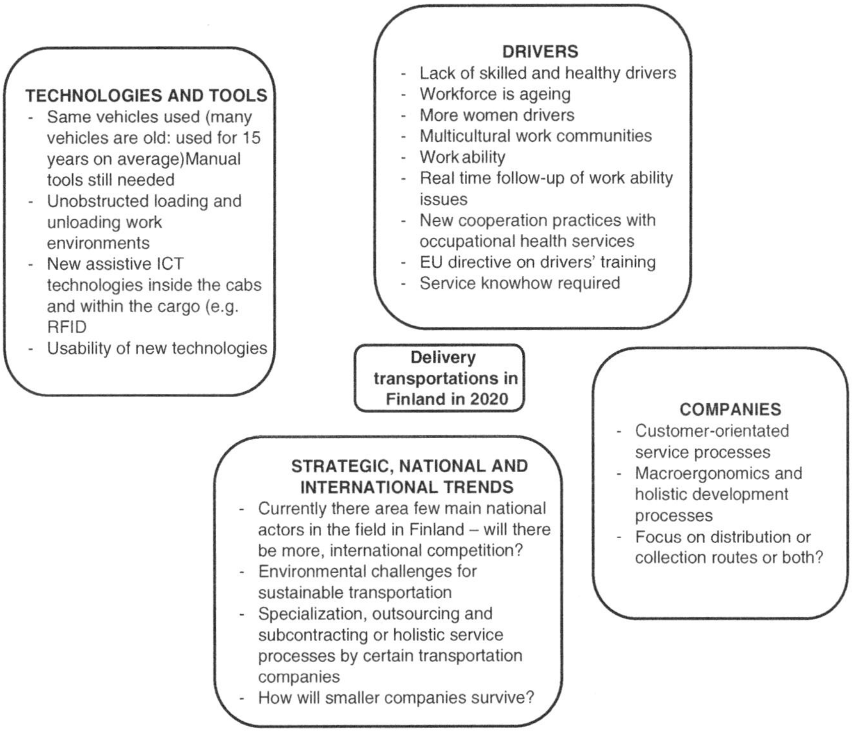

TECHNOLOGIES AND TOOLS
- Same vehicles used (many vehicles are old: used for 15 years on average) Manual tools still needed
- Unobstructed loading and unloading work environments
- New assistive ICT technologies inside the cabs and within the cargo (e.g. RFID)
- Usability of new technologies

DRIVERS
- Lack of skilled and healthy drivers
- Workforce is ageing
- More women drivers
- Multicultural work communities
- Work ability
- Real time follow-up of work ability issues
- New cooperation practices with occupational health services
- EU directive on drivers' training
- Service knowhow required

Delivery transportations in Finland in 2020

STRATEGIC, NATIONAL AND INTERNATIONAL TRENDS
- Currently there area few main national actors in the field in Finland – will there be more, international competition?
- Environmental challenges for sustainable transportation
- Specialization, outsourcing and subcontracting or holistic service processes by certain transportation companies
- How will smaller companies survive?

COMPANIES
- Customer-orientated service processes
- Macroergonomics and holistic development processes
- Focus on distribution or collection routes or both?

Fig. 5 Summary of future scenarios provided by focus group session

4 Discussion

The results of this study show, as expected, that L/SH drivers' work contains various risks of accidents and musculoskeletal disorders. However, they also show how the participatory approach can be utilized to find system-level development possibilities for the future.

According to the statistics analysis, over half of the accidents in the industry occur while the driver is performing tasks related to manual material handling, such as handling cargo and lifting or carrying different objects. The types of accidents in manual material handling tasks are rather mixed and may typically include slips, trips and falls or sudden overexertion or straining while carrying objects. The results of this study provide in-depth knowledge on accident risks and strengthen existing knowledge on the subject (see e.g. McClay 2008; Olson et al. 2009; Shibuya et al. 2010).

Regardless of the location in which the work is performed, employers are responsible for ensuring safe working conditions for their employees. Several kinds of risks exist in the work environments in which drivers perform work outside the cab. A majority of these are regular threats that can be responded to in a systemic manner. Nonetheless, as noted by Westrum (2006), irregular threats and unexampled events may also occur in work. Hollnagel (2008) emphasizes companies' and individuals ability to react to these changes; i.e. their resilience. In practice, in cases such as L/SH operations, in which employers have only restricted possibilities to supervise employees at work, they usually place their trust in the employees' ability to perform work safely in every oncoming situation, as Hale and Borys (2013) point out. As well as drivers' own personal skills, abilities and attitudes towards safety, organizational processes and safety culture must also be emphasized and developed continuously.

The video observation processes and subject population sizes, and the outcomes from different perspectives have been analysed in depth and discussed more profoundly earlier by Reiman (2013) in his Ph.D. thesis. The results of both Reiman (2013) and this study inevitably show that improving L/SH drivers' work in environments other than truck cabs requires the participation and continuous cooperation of the relevant stakeholders, through strategic short-term and long-term goals.

The development work in the L/SH sector should move from partial optimization towards global impacts, and system-level development processes. Sustainability is also an issue to consider in the field of logistics. Considerable actions and steps towards sustainability have been taken both nationally and internationally. Nonetheless, as emphasized by Black (2010) and Schiller et al. (2010), current transportation systems can still be considered non-sustainable. Zink and Fischer (2013) point out that ergonomics is one framework for improving sustainability from the human perspective. A premise in ergonomics is that employees can perform their work using safe work equipment in a safe work environment. A study by Shevtshenko et al. (2012) simplifies how sustainable design and service-orientated products that meet end-users' requirements, as in their case, fork lift trucks, both improve human performance and lead to lower lifetime costs than when the work is performed using conventionally designed products. Work environment and work equipment design issues need to be considered more carefully, as women are increasingly being employed as truck drivers. In this sense, Grandjean (1988) notes that anthropometric body size dimensions, which vary widely between individuals, between the two sexes and different races, and also between younger and older people, are important factors to be acknowledged in the trucking industry.

As Wioland (2013) points out, road transportation is one essential part of the economic performance of a country. Nonetheless, even though L/SH operations comprise the largest segment of the trucking industry (see e.g. Hanowski 2000) it can be roughly agreed that in Finland it lacks a holistic innovation system. This is mainly due to the size of the companies, but also to a lack of communication and cooperation in the value chain. The goal in the future is to find and connect strategic cooperation processes in the L/SH sector. Public authorities must be included in these processes, as they enable system-level innovations in addition to linear and often isolated product, service and process development activities by single companies. The underlying idea is that the future of competitive advantage is in structures and systems, bringing together manufacturers, suppliers, importers, transport operators, marketers, retailers, customers and other stakeholders. This development requires a holistic, resilient perspective.

The 2020 scenarios illustrated a future world in which technology enables a number of new things. It will be possible to measure drivers' health and performance if necessary; anywhere and anytime. User-driven information systems will provide information and advice for drivers on their work and even anticipate changes in circumstances (for example, possible distractions). A competent, trained driver will be assisted by a "virtual assistant" as part of the interactive transport operations control system. The control system will be able to instruct, for example, how to act at a particular loading or unloading place at the moment the driver arrives. This will be possible because different system operators can communicate with and understand each other. This system will reduce waiting times and provide high- quality service to the service chain for all parties involved. However, despite these different intelligent aids and tools being possibly available, the work will still contain physical activities. Thus drivers' health and work ability will remain an area in which to continue development work.

5 Conclusions

L/SH operations contain various work tasks that are performed by drivers in many work environments other than the truck cab. The systemic three time perspective approach (past, present and future) shows that the quality of the work environment depends on many stakeholders' actions, on both the micro- and macroergonomic level. Current design and management processes lack cooperation and communication practices. Thus, single transportation companies have only restricted possibilities for holistic work system management in the different work environments in which the work is performed. Although future scenarios suggest that new technologies will be available to ease L/SH operations, tasks will still require physical work on the part of the driver. Video analyses provide important, in-depth information on such work tasks. Illustrative video analyses, combined with concrete statistical analyses of the expenses of adverse health outcomes, can be utilized to justify the needs for micro- and macroergonomic improvements at the company level, and furthermore in strategic system-level issues, at the broader, macroergonomic stakeholder level.

Acknowledgments The authors would like to express their gratitude to the Finnish Work Environment Fund, the Finnish Funding Agency for Technology and Innovation, the Ministry of Social Affairs and Health, the Tauno Tönning Foundation, the Auramo Foundation and the companies and drivers who participated in the data collection and analyses, as well as to all other individuals and organizations who were involved in this development project over the last years.

References

Bisantz AM, Drury CG (2005) Applications of archival and observational data. In: Wilson JR, Corlett EN (eds) Evaluation of human work, 3rd edn. Taylor & Francis, Boca Raton, pp 61–82

Black WR (2010) Sustainable transport: problems and solutions. The Guilford Press, New York

Brown O Jr (2002) Macroergonomics methods: participation. In: Hendrick HW, Kleiner BM (eds) Macroergonomics. Theory, methods and applications. Lawrence Erlbaum Associates Inc, Mahwah, pp 25–44

Carayon P (2009) The balance theory and the work system model…twenty years later. Int J Hum Comput Interact 25(5):313–327

Dul J, Neumann PW (2009) Ergonomics contributions to company strategies. Appl Ergon 40(8):745–752

EN ISO 6385 (2004) Ergonomic principles in the design of work systems. European Committee for Standardization

European Agency for Safety and Health at Work (2010) A review of accidents and injuries to road transport drivers. Publications Office of the European Union, Luxembourg

European Agency for Safety and Health at Work (2012) Worker participation practices: a review of EU-OSHA case studies. Literature review. Publications Office of the European Union, Luxembourg

Forsman M (2008) Participatory ergonomics—supporting tools. In: Kujala J, Iskanius P (eds) Proceedings of the ICPQR conference. Oulu, pp 195–204

García Acosta G, Lange Morales K (2008) Macroergonomic study of food sector company distribution centres. Appl Ergon 39(4):439–449

Grandjean E (1988) Fitting the task to the man. A textbook of occupational ergonomics, 4th edn. Taylor & Francis, London

Gyi D, Sang K, Haslam C (2013) Participatory ergonomics: co-developing interventions to reduce the risk of musculoskeletal symptoms in business drivers. Ergonomics 56(1):45–58

Hale A, Borys D (2013) Working to rule or working safely? Part 1: a state of the art review. Saf Sci 55:207–221

Hanowski RJ (2000) The impact of local/short haul operations on driver fatigue. Dissertation, Virginia Polytechnic Institute and State University

Hanowski RJ, Wierwille WW, Garness SA, Dingus TA, Knipling RR, Carroll RJ (2000) A field evaluation of safety issues in local/short haul trucking. In: Proceedings of the human factors and ergonomics society annual meeting, pp 365–368

Hendrick HW (2002) An overview of macroergonomics. In: Hendrick HW, Kleiner BM (eds) Macroergonomics. Theory, methods and applications. Lawrence Erlbaum Associates Inc., Mahwah, pp 1–24

Hendrick HW, Kleiner BM (2001) Macroergonomics—an introduction to work system design. Human Factors and Ergonomics Society, Santa Monica

Hendrick HW (2007) A historical perspective and overview of macroergonomics. In: Carayon P (ed) Handbook of human factors and ergonomics in health care and patient safety. Lawrence Erlbaum Associates, Mahwah, pp 41–60

Hevner AR, Chatterjee S (2010) Design research in information systems, integrated series in information systems, vol 22. Springer, Heidelberg

Hollnagel E (2008) Risk + barriers = safety? Saf Sci 46(2):221–229

Huang YH, Zohar D, Robertson MM, Garabet A, Lee J, Murphy LA (2013) Development and validation of safety climate scales for lone workers using truck drivers as exemplar. Transp Res F Traffic Psychol Behav 17:5–19

Järvinen P (2007) Action research is similar to design science. Qual Quant 41(1):37–54

Kadefors R, Forsman M (2000) Ergonomics evaluation of complex work: A participative approach employing video-computer interaction, exemplified in a study of order picking. Int J Ind Ergon 25(4):435–445

Karwowski W (2005) Ergonomics and human factors: the paradigms for science, engineering, design, technology and management of human-compatible systems. Ergonomics 48(5):436–463

Keyserling WM, Monroe KA, Woolley CB, Ulin SS (1999) Ergonomic considerations in trucking operations: an evaluation of hand truck and ramps. Am Ind Hyg Assoc J 60(1):22–31

Kleiner BM (2006) Macroergonomics. In: Karwowski W (ed) International encyclopedia of ergonomics and human factors. CRC Press, Boca Raton, pp 154–156

Kleiner BM (2008) Macroergonomics: work system analysis and design. Hum Factors 50(3):461–467

Kraemer S, Carayon P, Sanquist TF (2009) Human and organizational factors in security screening and inspection systems: conceptual framework and key research needs. Cogn Technol Work 11(1):29–41

Kuorinka I (1997) Tools and means of implementing participatory ergonomics. Int J Ind Ergon 19(4):267–270

Lindhult M (2008) Are partnerships innovative? In: Svensson L, Nilsson B (eds) Partnership as a strategy for social innovation and sustainable change. Santerus Academic Press, Stockholm, pp 37–54

McClay RE (2008) Truck falls: examining the nature of the problem. Prof Saf 53(5):26–35

Miilunpalo P, Olkkonen S (2013) Kuljetus ja varastointi. In: Kauppinen T, Mattila-Holappa P, Perkiö-Mäkelä M, Saalo A, Toikkanen J, Tuomivaara S, Uuksulainen S, Viluksela M, Virtanen S (eds) Työ ja terveys Suomessa 2012. Seurantatietoa työoloista ja työhyvinvoinnista. Finnish Institute of Occupational Health, Helsinki, pp 191–196

Morel G, Amalberti R, Chauvin C (2009) How good micro/macro ergonomics may improve resilience, but not necessarily safety. Saf Sci 47(2):285–294

Olson R, Hahn DI, Buckert A (2009) Predictors of severe trunk postures among short-haul truck drivers during non-driving tasks: an exploratory investigation involving video-assessment and driver behavioural self-monitoring. Ergonomics 52(6):702–707

Pahl G, Beitz W (1988) Engineering design. A systematic approach. Springer, Berlin

Reiman A (2013) Holistic work system design and management—a participatory development approach to delivery truck drivers' work outside the cab. Dissertation, University of Oulu

Reiman T, Rollenhagen C (2014) Does the concept of safety culture help or hinder systems thinking in safety. Accid Anal Prev 68:5–15

Risikko T (2009) Safety, health and productivity of cold work. A management model, implementation and effects. Ph.D. thesis, dissertation, University of Oulu

Schiller PL, Bruun E, Kenworthy JL (2010) An introduction to sustainable transportation: policy, planning and implementation. The Cromwell Press Group, Bingley

Shevtshenko E, Bashkite V, Maleki M, Wang Y (2012) Sustainable design of material handling equipment: a win-win approach for manufacturers and customers. Mechanika 18(5):561–568

Shibuya H, Cleal B, Kines P (2010) Hazard scenarios of truck drivers' occupational accidents on and around trucks during loading and unloading. Accid Anal Prev 42(1):19–29

Smith M, Carayon-Sainfort P (1989) A balance theory of job design for stress reduction. Int J Ind Ergon 4(1):67–69

Smith CK, Williams J (2014) Work related injuries in Washington State's trucking industry, by industry sector and occupation. Accid Anal Prev 65(April):63–71

Strauss A, Corbin J (1998) Basics of qualitative research: techniques and procedures for developing grounded theory. SAGE Publications Inc, Thousand Oaks

Takala E-P, Pehkonen I, Forsman M, Hansson G-Å, Mathiassen SE, Neumann WP, Sjøgaard G, Veiersted KB, Westgaard R, Winkel J (2010) Systematic evaluation of observational methods assessing biomechanical exposures at work—a review. Scand J Work Environ Health 36(1):3–24

Ulrich KT, Eppinger SD (2004) Product design and development, 3rd edn. McGraw-Hill, Boston

van der Beek AJ (2012) World at work: truck drivers. Occup Environ Med 69(4):291–295

Vink P, Imada AS, Zink KJ (2008) Defining stakeholder involvement in participatory design processes. Appl Ergon 39(4):519–526

Westrum R (2006) A typology of resilience situations. In: Hollnagel E, Woods DD, Leveson NG (eds) Resilience engineering: concepts and precepts. Ashgate Publishing Limited, Aldershot, pp 55–66

Wioland L (2013) Ergonomic analyses within the French transport and logistics sector: first steps towards a new "act elsewhere" prevention approach. Accid Anal Prev 59:213–220

Zink KJ, Fischer K (2013) Do we need sustainability as a new approach in human factors and ergonomics. Ergonomics 56(3):348–356

Chapter 8
Fatal Non-driving Accidents in Road Transport of Hazardous Liquids: Cases with Review on Finnish Procedure for Investigating Serious Accidents Within Work System

Henri Jounila and Arto Reiman

Abstract The aim of this study was to review the Finnish investigation procedure for fatal accidents and to analyse four actual fatal accidents related to the transport of hazardous liquids, especially to service operations after the unloading phase. Content analysis was applied to identify the accident factors from the investigation reports. The accident factors were classified into two main classes, in which the factors related to safety culture and safety attitudes were discussed. The accident factors were also considered according to five elements of the work system model. Finally, as in the Finnish investigation procedure, the prevention measures of accidents were presented in condensed form.

1 Introduction

In Finland, fatal workplace accidents have declined in recent years. This can be considered relatively good, as fatal accident figures for Finland have previously been one of the lowest in country comparisons (see e.g., European Commission 2006; Health and Safety Executive 2014; Takala 1999). According to the preliminary estimate of the Federation of Accident Insurance Institutions (FAII), in Finland 20 workers died in workplace accidents in 2013 (Federation of Accident Insurance Institutions 2014b), while the figures in 2011 and 2012 were 28 and 26,

---

H. Jounila (✉)
Faculty of Technology, Industrial Engineering and Management, University of Oulu,
P.O. Box 4610, 90014 Oulu, Finland
e-mail: henri.jounila@oulu.fi

A. Reiman
Finnish Institute of Occupational Health, Aapistie 1, 90220 Oulu, Finland
e-mail: arto.reiman@ttl.fi

respectively (Federation of Accident Insurance Institutions 2013). For example, when the risk of death at work is discussed in the literature, the riskiest industry was that of transportation and storage in 2011 (Official Statistics of Finland [OSF]: Occupational accident statistics 2011).

Transport workers run up against numerous risks on the job. For example, physical work factors, such as painful positions, carrying or moving heavy loads, noise, vibration and high or low temperatures are a large part of numerous transport workers' workday. In addition, lack of sufficient training and work organisational factors such as repetitive work and non-standard working hours are common in the transport sector. Transport workers may also be exposed to dangerous chemicals in various service, maintenance and cleaning tasks (European Agency for Safety and Health at Work 2011c). According to studies related to truck drivers, many of the above-mentioned risk factors are also present in their work (see e.g., Okunribido 2006; Reiman et al. 2014). Various studies have paid special attention to non-traffic occupational risk factors or accidents. Reiman et al. (2014) identified both physical and psychosocial discomforts and risks associated with delivery truck drivers' work outside their cabs, while falls from heights were the main accident type in the truck drivers' loading and unloading work in Shibuya's et al. (2010) study.

Today, when we consider accidents at work, the terms safety culture and safety attitudes often emerge. The definition of safety culture has been examined in several studies, which demonstrate that the safety culture concept was born after the Chernobyl accident (Cooper 2000; European Agency for Safety and Health at Work 2011b; Reiman and Rollenhagen 2014; Ruuhilehto and Vilppola 2000). After that accident, the International Nuclear Safety Advisory Group defined the safety culture concept and emphasised in their definition an important role for organisations and individuals in safety issues (INSAG 1991, p. 1):

> Safety culture is that assembly of characteristics and attitudes in organizations and individuals which establishes that, as an overriding priority, nuclear plant safety issues receive the attention warranted by their significance.

In Reiman and Rollenhagen's (2014) study of the literature on the concept of safety culture, they found that behaviour, values, attitudes and beliefs were typically discussed. In particular, safety attitudes and their significance as a part of the safety culture appear in many research articles (see e.g., Cheyne et al. 2002; Harvey et al. 2001). Poor safety attitudes and safety culture (or safety climate) have been proved to play a considerable role in discussions of accidents at work (see e.g., Gillen et al. 2002; Zohar 2000). However, Reiman and Rollenhagen (2014) note that safety culture is only the starting point of the accident investigation, and that the investigation should not end at that point.

In the accident investigation, several accident factors are generally found; at both the organisation level and the individual level (see e.g., Jounila and Reiman 2012; Tihinen and Ijäs 2010). Cantor (2008) states that there has been increased interest in the research of the accident factors contributing to workplace accidents. It is important that the investigation is not finished when immediate accident factors have been found; rather, the investigation should continue until the root causes are

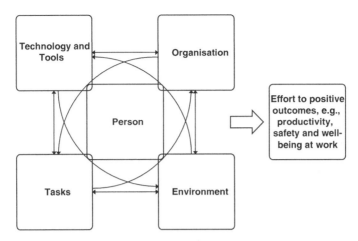

Fig. 1 The work system model (modified from Smith and Carayon-Sainfort 1989; Carayon 2009)

found. Both the immediate causes of the accident and the underlying causes must be considered (Kletz 1994). The accident factors and other results discovered in the accident investigation have to be entered into the investigation report, through which the investigation information can be distributed to workplaces and workers doing similar work. Lind (2009) also states that the accident reports include essential information on the accidents and suggest how these accidents can be avoided in the future, but companies do not utilise the reports to a sufficient extent. All work communities should have to learn about the accidents. The employer shall ensure that all elements of the work are in order, and that the employees can work safely and healthily every day.

The work system model (Fig. 1) is one way to describe the elements of work as a whole. One focus is a person, or worker, who produces wanted results. The elements include interaction between each other. In addition to a person, other elements of the work system model are tasks, technology and tools, organisation and environment. When a person is working on his/her tasks, he/she needs certain tools (technology) and works in some organisation and physical environment. When a change occurs in one part of the model, it also causes changes in other elements of the model (Carayon 2009). Finding "the equilibrium" is very challenging. The complexity of the work system model has been examined in various articles (see Autio et al. 2011; Carayon 2009; Carayon and Smith 2000); it has been, for example, stated that a work system produces psychosocial, cognitive and physical loads, which lead, in best-case scenarios, to positive outcomes, such as workers' safety and well-being at work (Carayon 2009). The work system model could be utilised in accident investigation processes to provide more in-depth data of the factors leading to the accident.

Finnish Procedure for Investigating Fatal Accidents at Work

According to the Finnish legislation, all insurance companies concerning the statutory accident insurance have to be members of the FAII. These companies are under an obligation to keep statistics on work accidents and to give out required information for maintaining national statistics. One of FAII's tasks is to investigate fatal accidents where an employed person die performing his/her working tasks in workplace (Tómasson et al. 2011). The fatal accident at work investigation report (TOT report) is produced on the basis of the investigation.

In addition to the FAII, police and, for example, occupational safety authorities carry out their own investigation about the fatal accidents (Fig. 2). FAII's TOT Board decides, according to their criteria, which fatal cases will be taken under investigation. The criteria are, for example, seriousness of the accident and frequency of similar accident. When it has been decided to investigate a certain case, the FAII expert is appointed to manage the investigation. He also acts as a secretary of the TOT investigation team (Federation of Accident Insurance Institutions 2012a).

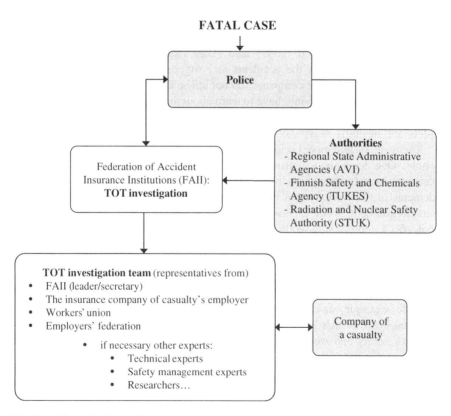

Fig. 2 Outline of the institutions that participate in the investigation of fatal accidents in Finland

The TOT investigation team is comprised of the required number of members. They have to have relevant experience in the field of the case in question, the working environment and the working practices. The team is comprised of experts from the insurance company, workers' union and employers' federation. In addition, the team might need technical and safety management experts, depending on the case (Fig. 1) (Federation of Accident Insurance Institutions 2012b).

During the early stages of investigation, the nominated investigation team strives to visit the accident site as soon as possible. Initial data (obtained from the victim's company, police and occupational safety authorities) is supplemented by discussions with those present at the time of the accident and the company's representatives. If necessary, the investigation team can request an expert opinion on the technical details, from some outside organisation.

The first versions of the TOT investigation report are drawn up on the basis of the gathered material and information (Fig. 3). The report is reviewed in the meetings of the investigation team and through information and communication technology (ICT). Finally, when the working group's last comments and corrective actions have been entered, the group accepts their own version of the written report. After this, the TOT investigation report will go to FAII's TOT Board for approval. When possible amendments and additions have been entered, the report is

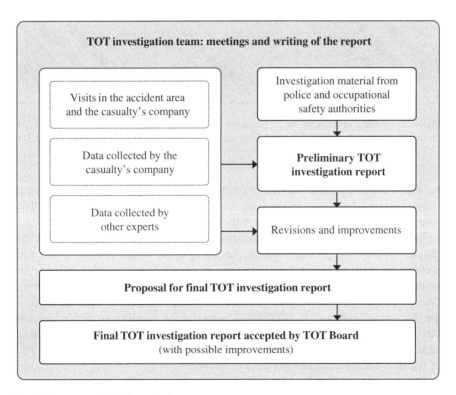

Fig. 3 Progress of TOT investigation

Fig. 4 The generic structure of the investigation report: the cover, text chapters with pictures and the course of events figure (the figure has been created from the TOT investigation report; see TOT 11/11 2012)

published and is available for utilisation by all working communities and by other people who are interested in the fatal accident in question. All investigated TOT cases are published in PDF format on the Internet at www.tvl.fi/totti.

The TOT investigation report consists of three main chapters (Fig. 4). The first chapter is the 'Background', where the background issues related to the fatal accident are reviewed, including the course of events from normal working to the moment of the accident, as well as the victim's work experience and other things that are considered necessary for understanding the events. The second chapter is the 'Factors of accident'. In it, the main factors of the accident are divided beneath their own subtitle, and each of the factors is briefly reviewed. The third chapter addresses the prevention of similar work accidents. It also consists of subtitles, each of which concerns particular prevention measure.

The report contains, in addition to the above-mentioned text chapters, a cover page with the case number and year of the accident, the report title, the abstract, the picture, and the caption. The report concludes with the table of ESAW codes of the case (see e.g., European Statistics on Accidents at Work 2012), the figures that illustrate the course of events, and the factors of the fatal accident.

2 Objectives

This study was made after the so-called theme investigation, in which the investigation team studied a fatal case which had occurred in 2011. In addition to that case, the investigation team analysed three cases which had occurred previously between the years 1993 and 2008 (Federation of Accident Insurance Institutions 2014a). Each of these three cases had been investigated previously, and a 'Fatal accident at work Investigation report' had been written for each of them. All of these four cases were synthesised by the TOT investigation team, which then produced the 'Theme investigation report'. In the report, the theme was the fatal accidents in service operations of the tanker after the unloading phase of hazardous liquids (especially during or after the washing phase of tanks).

After the theme investigation, the first part of a further study was carried out by Jounila and Reiman (2012). The main goal of that study was to find out whether there were any similar features between the investigated accidents. The main emphasis was on the factors that led to the fatal accidents. A few findings emerged and these will be taken into account in applicable parts in this study.

This study was carried out because deeper information was needed about the fatal cases, especially information about the factors related to the safety attitudes and safety culture of the organisations. This information might be utilised in safety management of transport companies as well as on the other side of the supply chain, such as in the decision making of purchasing enterprises which buy transport services. Many other investigation teams in Finland have carried out theme investigations of other areas, for example, areas related to transferable ladders. Thus it is highly significant to gain experimental information about the used method.

In summary, two aims were set for this study:

- Aim 1: to study safety attitude- and safety culture-related factors that have contributed to the accidents in the cases
- Aim 2: to study in the work system framework all the accident factors that were found in the examination

3 Methods

This study consisted of the following parts (Fig. 5):

- exploration of existing theory, including journal articles and other relevant literature

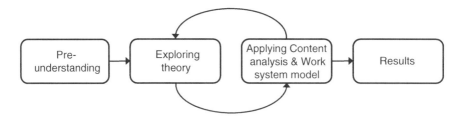

Fig. 5 The research process

- the review of the Finnish TOT investigation method, based on the personal experiences of the TOT investigation team, long-time utilisation of TOT investigation reports in university teaching, participation in several expert lectures, and exploration of the TOT investigation reports and other available documents and information on the Internet
- study of four fatal cases related to transport of hazardous liquids in Finland, applying content analysis and work system framework (more at the end of this chapter)

In this study, four existing 'fatal accident at work investigation reports' were used. These are generally available on the Internet by FAII (Federation of Accident Insurance Institutions 2014a). All these four documents are related to transportation of hazardous liquids, and to work tasks which have been carried out after the unloading operation.

In analysing the TOT investigation reports, content analysis was applied (Tuomi and Sarajärvi 2009) and a work system framework was set up (see Introduction). The study is qualitative in nature, but it also incorporates quantitative features. Various steps were conducted based on the analysis method, the contents of the TOT investigation method and the aims of this study:

1. compilation of material (TOT investigation reports)
2. in-depth reading of the reports (previous examination of the reports facilitated this phase)
3. in pursuit of text reading, reflection on the text and preliminary coding (word(s), concept or clauses) and preliminary creation of the classification units (previous experience was utilised in the creation of the classes)
4. final coding and final selection of classification units
5. separation of the classes related to safety attitudes and safety culture into their own groups
6. placement of classes into the work system model
7. discussion and conclusions

4 Results

In this study, several accident factors (classes) were identified (see Table 1). The factors were separated into two main groups: (1) basic information and (2) information related to safety attitudes and safety cultures. The first one included 11 classes and the latter one 8 classes. Three drivers and one cleaning company's mechanic died in these four accidents at work. Two victims worked on the tank and two inside the tank. Two workers' cause of death (affected factor) was toxic vapours of raw turpentine and the other two lack of oxygen (for more details, see more Table 1 and the investigation reports) inside the tank. The time of day was, in three cases, probably in the evening, whereas in one case the time is not clear from the investigation report. The rest of the basic information classes created are in the Table 1.

Another main group is comprised of the information related to safety attitudes and safety culture. The work experience of victims was mainly considered sufficient. Instead, there were serious deficiencies in the use of personal protective equipment (PPE). Respirators were not in use; there were not available at all or they were improper for the working conditions in which workers used them. Other workers noticed the possible accident quite soon in two cases and after a while in one accident (when they returned from the break room). One driver was working alone.

Received training, job orientation and instructions differed very widely among the victims. Many good practices were identified, but there were a lot of improvable subjects in these areas (see details in Table 1). In all cases, some improper actions were identified; for example, respirators were not used and the tanks were washed in incorrect places. In at least two cases, there was some confusion between the organisations; they had not explicitly agreed on unloading and washing, supervision and responsibilities, or information was not communicated. Supervision of work was inadequate in most cases and some confusion between organisations was also identified.

The accident factors were grouped according to five elements of the work system model (Fig. 6). The original aim was to place each of accident factors only under one element; that is to say, the factor best suited under that element. Only two factors had to be placed in two elements: 'specially mentioned improper actions' was placed in the task and the person elements and 'training, job orientation, instructions etc.' was placed in the person and the organisation elements. It was not possible to place them in one element when they belonged clearly and equally in two elements. The environment element included only basic information (see Table 1); the factors related to chemicals were also placed in the environment element. As expected, the person and the organisation elements received most factors related to safety attitudes and safety culture while the technology and tools element received one less.

Table 1 Contents of the investigation reports (TOT no/year) were divided into 2 main groups and 19 classes (updated and modified from Jounila and Reiman 2012)

Case number/year	TOT 4/11	TOT 4/08	TOT 13/96	TOT 1/93
Basic information				
(Profession of) victim	Driver	Driver	Driver	Mechanic (tank-cleaning company's worker)
Age	49	39	57	N/A
Accident location	Mill site	Transport company	Mill site	Tank-cleaning company
Exact (accident) site	Washing place	Washing hall	Unloading place	Service hall
Worker's position	On the tank	On the tank	In the tank	In the tank
Outside/inside	Outside	Inside	Outside	Inside
Discharged liquid	Raw turpentine	Raw turpentine	Sulphuric acid	Lubricating oil (by using nitrogen pressure)
Cause of death (affected factor)	Toxic vapours	Toxic vapours	Not enough oxygen in the lowest point of the tank (high sulphuric acid level)	No oxygen in the tank (high nitrogen level)
Task before accident	Unloading	Unloading	Unloading	Unloading
Task when accident happened	Tank cleaning	Tank cleaning	Tank maintenance after tank cleaning	Tank cleaning
Time when accident happened	Late in the evening (at night)	Early evening	Evening	N/A
Information related to safety attitudes and safety culture				
Experience	25 years (transporting a variety of materials, e.g., raw turpentine)	At least 3 years as a tank truck driver. The employer: adequate experience	Very experienced/skilled	1.5 years in this company. The worker had washed dozens of tanks

(continued)

Table 1 (continued)

Information on personal protective equipment PPE	Only a helmet with a visor (in use)	Respirator in the vehicle (not in use; inadequate respirator in this case)	No respirator in use nor in the vehicle	Respirator with activated carbon filter (in use) (improper protective equipment at low oxygen concentrations)
Other workers	All alone	Workers in the garage sensed a strong smell of raw turpentine so they exited to the break room	A control room worker saw that the driver descended into the tank. He went to check because the driver did not return	Two other workers descended into the tank to rescue mechanic and lost consciousness. However, these two survived
Difficulties	Perhaps failure in the drain pump for washing water		Faulty valve on the bottom of the tank	
Training, job orientation, instructions, etc.	The mill site (in question) provided safety training a year earlier. The company's experienced drivers give job orientation to drivers. Work in the tank-cleaning place had been gone through between the companies. According to the contract, the drivers' employer takes charge of work-related arrangements, job orientation and preparation of work instructions	Hazardous material driver's license (2004). The employer's safety instructions (learned and confirmed by signature). The employer: the job orientation was OK. The driver received information about the dangers of raw turpentine by message and also at the place of loading (written safety instructions in his native language). However, insufficient instructions and communication. The conveying of tacit knowledge was inadequate	The driver had been given instruction on tank work. The company did not have any written instructions on going into the tank, measuring in the tank or safety actions in this kind of case	The company's employees had acquired training for the oil tank clean-up work. oral Instructions had been received from drivers, for example

(continued)

Table 1 (continued)

Specifically mentioned improper actions	Respirator not used	Respirator not used. Cleaning operations should be carried out in the mill site, in the open air and immediately after discharging. Washing a raw turpentine tank was forbidden in the washing hall (whether the driver knew this is not known)	Tanks must not be washed at the mill site. The driver did not ask permission for washing	Because the risk was not recognised, the wrong respirator was used, the tank was not ventilated and measured, another person did not monitor outside the tank, and the rescue operation was started with inadequate PPEs
Confusion among organisations		The organisations had not explicitly agreed on unloading and washing, supervision and responsibilities (between the factory, the transportation company and the subcontractor who eventually took care of transportation)		Very experienced driver did not receive any information about the use of nitrogen in the unloading phase (from the orderer's worker). Thus, the cleaner did not receive any either. In the consignment note, there was no note about the use of nitrogen. The driver believed that air pressure was used. If the cleaner had known about the use of nitrogen, the working method would have been different
Supervision of work	No supervision. The driver was found dead early in the morning	Inadequate supervision. The supervision of the washing did not belong to the mill workers. The transport organiser was not able to control an individual driver	The mill worker monitored working from the control room	Another person did not monitor outside the tank

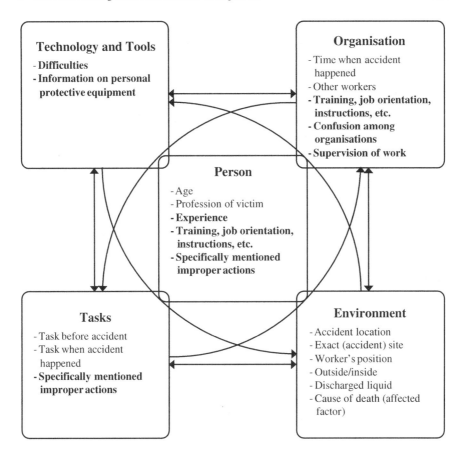

Fig. 6 The accident factors (see also Table 1) in the work system model. Information related to safety attitudes and safety culture is in boldface

5 Discussion

Several accident factors were identified from the investigation reports discussed. Most of the factors were connected to safety culture and safety attitudes one way or the other, but still part of the factors were eliminated; only a connection that clearly occurred enabled entry into this category, and the others were transferred to 'basic information'. On the basis of the investigation reports, it is clear that special attention should be paid to the use of personal protective equipment in tank-cleaning tasks. Employers have to supervise the use of PPEs as well as all other work. In addition to the use of PPE, a few other issues were identified, related very clearly to safety attitudes (and safety culture): for example, washing rules were not obeyed and the verifying worker did not monitor outside the tank. Supervision of work and emphasising safety working conditions are employers' continuous work.

They must require and monitor the following the rules. Another worker can verify safety, working as a twosome. That is especially important if the insides of the tanks are being worked on.

As mentioned earlier, received training, job orientation, instructions, etc. varied among the workers, at least on the basis of the investigation reports. However, it is good to notice that these accidents occurred between the years 1993 and 2011, and therefore a lot of improvements have been made on these issues in work communities and society during these years. This can be also noted in Table 1, where information is clearly less available in the oldest cases. In any event, improvements could have been made in all cases in relation to the above-mentioned issues. As an example of this area, the Driver's manual was issued in a company in Finland (European Agency for Safety and Health at Work 2011a). The manual includes important information for company's drivers, such as responsibilities and safety matters at work and harmful and dangerous substances. The Driver's manual has also been utilised for the educational and job orientation purposes of the drivers.

Today, numerous different organisations are connected with many single work tasks. As one example, a driver of the subcontracting company picks up the load from the one mill site and transports it to the other one. After this, he takes the containers to the tank-cleaning company. In this case, the supply chain already includes five organisations: the actual transportation company and its subcontractor, two mills and the tank-cleaning company. It is extremely important that the contracts between these companies are made explicit and that information flows between organisations. Special attention must be paid to the supervision of work and to responsibilities.

The organisation has a very important role in the work system as a factor contributing to safety (see Autio et al. 2011). The organisation, most often the employer, provides work tasks and is therefore responsible for ensuring that employees are able to do their jobs safely and healthily. The employer is required to assess the risks of the work and to supervise employees' work. The organisation is also responsible for the fact that things are going on between organisations. Another significant factor in the work system is the person (see Autio et al. 2011), the worker, who must follow the instructions. The employee must have sufficient experience of the work and the dangers of the work. The organisation shall ensure that the employee has sufficient competence, and that every worker receives a comprehensive job orientation and instructions for work.

Content analysis was utilised in this research, which proved usable in this kind of study. The method enables processing material (the fatal investigation reports) comprehensively and in depth. The work system framework seems to be suitable for studies of a various kinds related to work. The accident factors were able to be placed in the elements of the work system model, but some difficulties occurred. Many of the accident factors, for example, could have been placed in various elements. However, the original plan was to place one accident factor in only one element; mainly, this was fulfilled. One significant addition to this study would have been to consider more explicitly the interrelationship of the accident factor for all five elements. It is clear that the number of investigation reports was limited, but

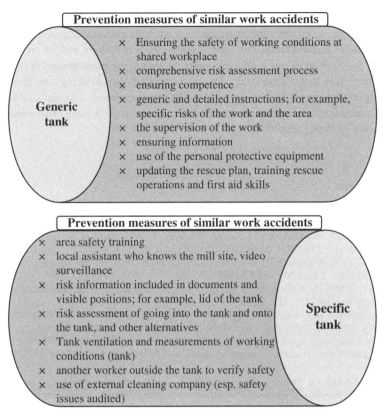

Fig. 7 Some generic and specific prevention measures of accidents compiled from the investigation reports, with a few additions. There are details and more prevention measures of accident in the investigation reports (Federation of Accident Insurance Institutions 2014a). It should be remembered that the risks of the prevention measures also have to be assessed separately in each work environment

more accident reports were not available in this area. Thus the generalisation of the results has to be done carefully. In future, the method used could be repeated with the broader material.

6 Conclusions

One of the most significant aims of the TOT investigation system is to inform work communities about the fatal accidents. The investigation report provides the important information for work places that operate in the same area or do the same kind of work as done in the investigated case. Thus the investigation and produced report aim for the prevention of similar work accidents. Therefore, it was important to compile the most significant prevention measures from these four investigation reports. In Fig. 7, the prevention measures have been presented as condensed. The Figure can be utilised in training, job orientation, instructions and so on. It is also suitable for tank work on applicable parts. It should be remembered that the risk assessment has to be performed for all work in those working conditions and working environments where the work tasks are performed. The risks of the prevention measures have also to be assessed separately in each work environment and in various working conditions.

References

Autio T, Jounila H, Sinisammal J, Reiman A (2011) Factors of well-being at work in the work system model. In: Savolainen M, Kropsu-Vehkaperä H, Aapaoja A, Kinnunen T, Kess P (eds) Proceedings of TIIM2011 conference, Oulu, Finland, 28–30 June 2011

Cantor R (2008) Workplace safety in the supply chain: a review of the literature and call for research. Int J Logist Manage 19(1):65–83. doi:10.1108/09574090810872604

Carayon P (2009) The balance theory and the work system model…twenty years later. Int J Hum Comput Interact 25(5):313–327

Carayon P, Smith M (2000) Work organization and ergonomics. Appl Ergon 31:649–662

Cheyne A, Oliver A, Tomás JM, Cox S (2002) The architecture of employee attitudes to safety in the manufacturing sector. Pers Rev 31(6):649–670. doi:10.1108/00483480210445953

Cooper MD (2000) Towards a model of safety culture. Saf Sci 36(2):111–136. doi:10.1016/S0925-7535(00)00035-7

European Agency for Safety and Health at Work (2011a) Managing risks to drivers in road transport. Working environment information: working paper. Publication office of the European Union, Luxembourg. https://osha.europa.eu/en/publications/reports/managing-risks-drivers_TEWE11002ENN. Accessed 14 June 2014

European Agency for Safety and Health at Work (2011b) Occupational safety and health culture assessment—a review of main approaches and selected tools. Working environment information: working paper. Publication office of the European Union, Luxembourg. https://osha.europa.eu/en/publications/reports/culture_assessment_soar_TEWE11005ENN. Accessed 9 June 2014

European Agency for Safety and Health at Work (2011c) OSH in figures: occupational safety and health in the transport sector—an overview. European risk observatory report. Publication office of the European Union, Luxembourg. https://osha.europa.eu/en/publications/reports/transport-sector_TERO10001ENC. Accessed 6 June 2014

European Commission (2006) Health and quality in work—final report, vol IV. Annex IV: Accidents at work. Annex V: Cross-sectional models. European Comission, Directorate General for Employment, Social Affairs and Equal Opportunities, Unit D1. http://www.ec.europa.eu/social/BlobServlet?docId=2133&langId=en. Accessed 6 June 2014

European Statistics on Accidents at Work (ESAW). Summary methodology (2012) 2012 edition. Eurostat, methodologies and working papers. Publications office of the European union, Luxembourg. http://epp.eurostat.ec.europa.eu/cache/ITY_OFFPUB/KS-RA-12-002/EN/KS-RA-12-002-EN.PDF. Accessed 5 June 2014

Federation of Accident Insurance Institutions (2012a) The TOT investigation procedure. http://www.tvl.fi/en/Investigation-of-accidents-at-work/The-TOT-investigation-procedure/. Accessed 30 May 2014

Federation of Accident Insurance Institutions (2012b) TOT-tutkintaryhmät. http://www.tvl.fi/fi/
   Tyopaikkaonnettomuuksien-tutkinta-TOT/TOT-tutkintaryhmat/. Accessed 30 May 2014
Federation of Accident Insurance Institutions (2013) Työtapaturmat – tilastojulkaisu 2013. http://
   www.tvl.fi/fi/Tilastot-/Tilastojulkaisut/Tilastojulkaisu/. Accessed 5 June 2014
Federation of Accident Insurance Institutions (2014a) (cases TOT 4/11, TOT 4/08, 13/96, 1/93).
   www.tvl.fi/totti. Accessed 17 April 2014
Federation of Accident Insurance Institutions (2014b) Ennakkoarvio 2013. http://www.tvl.fi/fi/
   Tilastot-/Tilastojulkaisut/Ennakkoarvio-2012/. Accessed 5 June 2014
Gillen M, Baltz D, Gassel M, Kirsch L, Vaccaro D (2002) Perceived safety climate, job demands,
   and co-worker support among union and non-union injured construction workers. J Saf Res 33
   (1):33–51. doi:10.1016/S0022-4375(02)00002-6
Harvey J, Bolam H, Gregory D, Erdos G (2001) The effectiveness of training to change safety
   culture and attitudes within a highly regulated environment. Pers Rev 30(6):615–636
Health and safety executive (2014) European comparisons. Summary of GB performance. http://
   www.hse.gov.uk/statistics/european/european-comparisons.pdf. Accessed 6 June 2014
International Nuclear Safety Advisory Group (INSAG) (1991) Safety culture, safety series no
   75-INSAG-4. http://www-pub.iaea.org/MTCD/publications/PDF/Pub882_web.pdf. Accessed
   9 June 2014
Jounila H, Reiman A (2012) Transport of hazardous liquids in Finland: an analysis of fatal
   accidents in service operations after the unloading phase. In: Antonsson A-B, Hägg GM (eds)
   Proceedings NES2012. Ergonomics for sustainability and growth. KTH Royal Institute of
   Technology, School of Technology and Health, Division of Ergonomics, Stockholm, Sweden
Kletz T (1994) Learning from accidents, 2nd edn. Butterworth-Heinemann Ltd, Oxford
Lind S (2009) Accident sources in industrial maintenance operations. Proposals for identification,
   modelling and management of accident risks. VTT publications 710. Dissertation, Tampere
   University of Technology
Official Statistics of Finland (OSF): Occupational Accident Statistics (2011) Wage and salary
   earners' accidents at work. Statistics Finland, Helsinki. http://www.stat.fi/til/ttap/2011/ttap_
   2011_2013-11-27_kat_001_en.html. Accessed 5 June 2014
Okunribido OO, Magnusson M, Pope M (2006) Delivery drivers and low-back pain: a study of the
   exposures to posture demands, manual materials handling and whole-body vibration. Int J Ind
   Ergon 36(3):265–273. doi:10.1016/j.ergon.2005.10.003
Reiman T, Rollenhagen C (2014) Does the concept of safety culture help or hinder systems
   thinking in safety? Accid Anal Prev 68:5–15. doi:10.1016/j.aap.2013.10.033
Reiman A, Putkonen A, Nevala N, Nyberg M, Väyrynen S, Forsman M (2014) Delivery truck
   drivers' work outside their cabs: ergonomic video analyses supplemented with national
   accident statistics. Hum Factors Ergon Manuf Serv Ind. doi:10.1002/hfm.20547
Ruuhilehto K, Vilppola K (2000) Turvallisuuskulttuuri ja turvallisuuden edistäminen yrityksessä.
   Tukes-julkaisu 1/2000. VTT Automaatio, Riskienhallinta, Helsinki. http://www.tukes.fi/
   Tiedostot/julkaisut/1-2000.pdf. Accessed 9 June 2014
Shibuya H, Cleal B, Kines P (2010) Hazard scenarios of truck drivers' occupational accidents on
   and around trucks during loading and unloading. Accid Anal Prev 42(1):19–29. doi:10.1016/j.
   aap.2009.06.026
Smith MJ, Carayon-Sainfort P (1989) A balance theory of job design for stress reduction. Int J Ind
   Ergon 4(1):67–79
Takala J (1999) Global estimates of fatal occupational accidents. Epidemiology 10:640–646
Tihinen T, Ijäs A (2010) Onnettomuustutkintaraportti, Dnro 11573/06/2010. Arizona Chemical
   Oy:n säiliöräjähdys 15.9.2010. Tukes. http://www.tukes.fi/Tiedostot/kemikaalit_kaasu/
   Onnettomuustutkintaraportti_Arizona_Chemical.pdf. Accessed 11 June 2014
Tómasson K, Gústafsson L, Christensen A, Solberg Røv A, Gravseth HM, Bloom K, Gröndahl L,
   Aaltonen M (2011) Fatal occupational accidents in the Nordic Countries 2003–2008.
   TemaNord 2011:501. Nordic Council of Ministers, Copenhagen

TOT 11/11 (2012) Paperityöntekijä jäi puristuksiin kartonkirullan ja telaston väliin. http://totti.tvl.
   fi/totcasepublic.view?action=caseReport&unid=906. Accessed 2 June 2014
Tuomi J, Sarajärvi A (2009) Laadullinen tutkimus ja sisällönanalyysi. Tammi, Helsinki
Zohar D (2000) A group-level model of safety climate: testing the effect of group climate on
   microaccidents in manufacturing jobs. J Appl Psychol 85(4):587–596. doi:10.1037//0021-
   9010.85.4.587

Chapter 9
The Total Risk of Lost-Time Accidents for Personnel of Two Large Employers in Finland

All Accidents Affecting Work: Home and Leisure, Commuting, and Work Injuries as Causes of Absence

Liisa Yrjämä-Huikuri and Seppo Väyrynen

Abstract This chapter presents a new kind of comprehensive data on all injurious accidents (10 years timespan) involving employees ($N = 13{,}000$) of two large workplaces in Finland. The main aim of this study was to clarify and assess the significance of risks among employees in different lost-time injury (LTI) accident categories, which were accidents at work, at home and during leisure time as well as when commuting to/from the job. Therefore, it was possible to reveal the total risk based on both frequency and absence time figures related to accidents. As far as the studied cases (a metal processing mill and a municipality) are concerned, home and leisure-time accidents appeared to be the most numerous category. Preliminary analysis of the metal processing case seemed to show that blue-collar employees have more numerous accidents of leisure in the same way they do have more numerous accidents at workplace, compared with white-collar employees.

Keywords Risk · Safety · Accident incidence rate · Injuries · LTI · Home and leisure accidents · Occupational accidents · Absence from work · Causes of absence · Off-the-job safety

---

L. Yrjämä-Huikuri · S. Väyrynen (✉)
Faculty of Technology, Industrial Engineering and Management, University of Oulu,
P.O. Box 4610, 90014 Oulu, Finland
e-mail: Seppo.Vayrynen@oulu.fi

L. Yrjämä-Huikuri
e-mail: Liisa.Yrjama@gmail.com

1 Introduction

According to one conclusion (Räsänen 2007), Finland represented the average rate in the EU as far as the incidence of accidents at work in 1994–1999. When all Nordic countries were compared during the same period, Sweden showed the lowest rates and Norway the highest, while Denmark and Finland were in between (Räsänen 2007).

As a whole, the Finnish Ministry of Social Affairs and Health concludes that the number of occupational accidents has clearly decreased during the last 20 years (Ministry of Social Affairs and Health 2005). The most problematic branches of industry have been manufacturing, buildings (construction), transport and the municipal sector (Räsänen 2007; The Federation of Accident Insurance Institutions 2007).

One of the key areas of development in Finland during the last years has been inter-organizational safety. Many employees often work at different so-called shared workplaces, and therefore, safety independent of the place where employees are, at each particular time, has been a goal (Väyrynen et al. 2008). But, how is the level of safety guaranteed when employees are not at any workplace? How are their safety aspects dealt with when we consider activities when commuting to/from the workplace and during leisure time and at home?

A nationwide interview survey (a statistically representative sample of 7,193 adult citizens) is a general source for finding out how people are involved in an accident or violence in Finland (Haikonen and Lounamaa 2010; Kansallinen uhritutkimus 2007). The different categories of accidents include: traffic, work, home, physical exercise, other accident (and violence). Finnish national official statistics concern mainly accidents at work (employed people, wage and salary earners) and traffic accidents, i.e. accident types covered by the obligatory insurance system. The question arises of what is the so-called total accident safety of Finns belonging to the personnel of typical workplaces. The nationwide official records do not show this. What types of accidents other than work-related ones happen to employed people, to what extent do they occur and what are their consequences? The survey mentioned earlier showed and analysed the diversity of categories of accidents in the Finnish adult population, generally.

The Swiss Council for Accident Prevention has also collected some detailed statistics (Sommer et al. 2007). The statistics cover all road, sports, household and leisure accidents among the resident Swiss population in 2003. One third were injured during sporting activities and almost 60 % of injuries happened at home and during leisure time.

It has been said by some companies that off-the-job safety is as important as on-the-job safety (Mottel et al. 1995); nevertheless, most work communities have not embraced that theory yet. On the other hand, e.g. sporting activities promote health but still incur losses for the national economy by way of sports accidents (Sommer et al. 2007). Accidents, happening at work or at home, create costs for employers also. Analysis has shown that off-the-job accidents are even more costly than lost-time on-the-job injuries, therefore it can be said that home and leisure safety would be good business also (Mottel et al. 1995).

An initiative of a metal processing mill, followed by interest of a big municipality provided an opportunity to take a comprehensive look at the 'total safety' of a sample of two large workplaces in Finland, with each of them having many thousands of employees.

2 Objectives

This study aimed to introduce in detail two representative cases by compiling statistics on all the accident data of personnel at a metal processing mill and a municipality, including work, commuting and home and leisure accidents (HLA). The amounts and proportions of accidents by each category, based on the total material and separately for the metal processing and municipality cases, are presented. Length of disability—during which an employee normally is absent from work—per accident and as total cumulative lost working time are analysed by category, based on the total material and separately for the metal processing and municipality.

3 Methods and Materials

The accident categories selected for this study were accidents at work (place) (AW), commuting accidents (CA), and HLA. At companies and public sector workplaces, like municipalities, AW and CA have to be registered and followed up on for legal and insurance reasons. Statistics on HLA are most often quite difficult to compile. In all, in the metal processing and municipality, solutions to this problem were found, though they both had a different solution. Occupational safety, accident insurance, and occupational health care functions were the contexts through which HLA data could be obtained for study purposes. The metal processing is different in that that the mill has for years utilised HLA as a routine because the mill has a special insurance system that includes HLA (Qvist et al. 2004; Qvist and Väyrynen 2005). Access to this kind of data is rarely possible at many workplaces. In practice, the results will hopefully provide inspiration and information on a more holistic safety culture and handling of accident risks in companies and at other workplaces as well as in society as a whole.

The metal processing and municipality comprise a total of around 11,800 injurious accidents, and the years covered by the data were, respectively, 2001–2013 and 2003–2011 (Qvist et al. 2004; Yrjämä and Ollanketo 2007; Yrjämä and Väyrynen 2006). The variables and their classifications were typically used in the occupational accident statistics, like length of disability (Kjellen 2000; The Federation of Accident Insurance Institutions 2007; The Statistics Finland 2007).

The length of disability, days of absence from work, describes the seriousness of the accident. A lost-time injury (LTI) was defined as an injury that causes at least one entire day to be lost from work. Some results also were analysed by using the LTI rate and Chi-square statistic.

The total number of employees in the two cases amounted to approximately 13,000. For an analysis, the employees of the metal processing were divided to the categories of production workers (blue-collar employees) and supervisors (white-collar employees). To facilitate comparison, the following data were also gathered:

- National level of AW and CA based on figures and variables of all injurious wage earners; accidents that happened and were reported to insurance companies within the Finnish industry and other workplaces during the corresponding years (The Federation of Accident Insurance Institutions 2007; The Statistics Finland 2007).
- A national survey could also be utilised (Kansallinen uhritutkimus 2007; Haikonen and Lounamaa 2010).

4 Results

Figures 1, 2, 3 and Table 1 show the basic results of metal processing by year, and accident category. Figures 4, 5 and Table 2 show the basic results of municipality. The remaining tables show more detailed analysis of metal processing.

If a rough estimation is made of how employees use their time in the same categories, a comparison can be made as to which activity category is most hazardous if incidence of injurious accidents is the comparison criterion. Employees are assumed to work 8 h/day, 5 days/week, 47 weeks/year on average. Every day

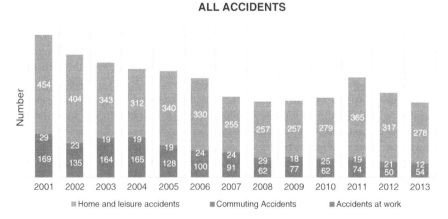

Fig. 1 The number of different accident types of metal processing by the year. The number of employees was 3,260 on average

Fig. 2 a–c Accidents (WA, CA, HLA) of personnel of metal processing, in the years 2001–2013. Calculated 95 % confidence interval (Upper/lower control limit)

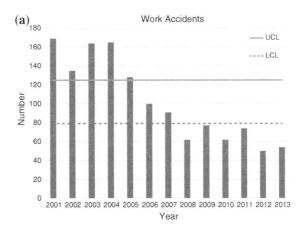
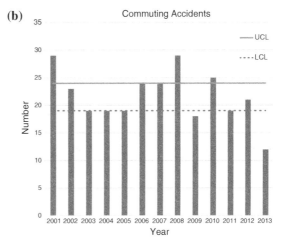
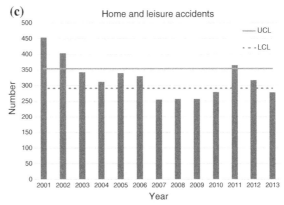

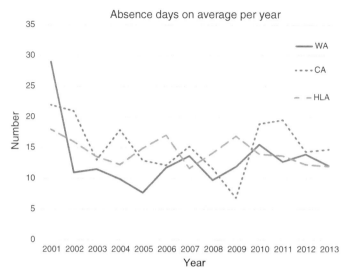

Fig. 3 Average absence days per one accident caused by different accident categories of metal processing, years 2001–2013

Table 1 Accidents of personnel and average absence time due to accidents (per one accident), including records from the years 2012–2013 of metal processing

Category of accidents of employees one absence day or more	Data concerning accident categories during each year					
	$n_{accident}$		Proportion (%)		Average absence time, days	
	2012	2013	2012	2013	2012	2013
AW	50	54	13	16	14	12
CA	21	12	5	3	14	15
HLA	317	278	82	81	12	12
Total	388	344	100	100		

The number of employees was 3,119 and 3,006, respectively

she/he sleeps 8 h. Every workday she/he needs to spend 1 hour commuting to and from the workplace. She/he uses the rest of her/his time for various home and leisure-time activities. The calculation leads to the following time-shares per each time category: workplace, 32 %; commuting to/from the workplace, 4 %; home and leisure-time activities, 64 %.

The LTI rate was estimated as the number of lost-time injurious accidents in various daily activities for people who are employed in a metal company (metal processing). The accidents of the last 10 years, from 2004 to 2013, were included. The base for the rate was one million hours of work (or commuting, or home and leisure):

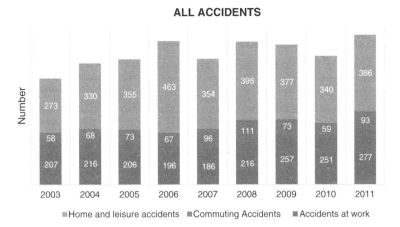

Fig. 4 Number of different accident types of municipality by the year (WA, HLA One or more sick leave days, CA Three or more sick leave days). The number of employees was 9,620 on average

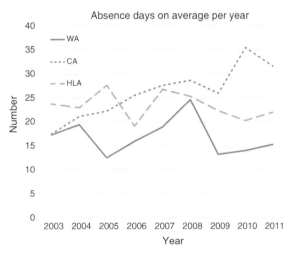

Fig. 5 Average absence days per one accident caused by different accident categories, of municipality, years 2003–2011 (CA includes cases with three absence days or more)

- LTI rate for WA (work place): 15.8
- LTI rate for CA (commuting): 24.3
- LTI rate for HLA (home and leisure): 25.6

Another way for comparing the accidental injury risk in activities during an average day the time-shares of main activities are known for an average employee. Contingency tables were collected, analysed and tested using the Chi-square statistic. For those data, the current years 2009–2013 were chosen from the metal processing company.

All accident categories were compared in a paired fashion. The contingency tables (cross tabulation), tables in a matrix format that displays the frequency distribution of the variables, were the following (Tables 3, 4 and 5). The analyses show that in the

Table 2 Accidents of personnel and average absence time due to accidents (per one accident), including records from the years 2010–2011 of municipality

Category of accidents of employees One absence day or more	Data concerning accident categories during each year					
	$n_{accident}$		Proportion (%)		Average absence time, days	
	2010	2011	2010	2011	2010	2011
AW	251	277	39	37	14	15
CA[a]	59	93	9	12	35	32
HLA	340	386	52	51	20	22
Total	650	756	100	100		

The number of employees was 9,709 and 10,043, respectively
[a] Three absence days or more

Table 3 Accident data (N) in the form of contingency tables, and the test results for comparing work place and commuting accidents

WA, actually	CA, actually
WA, if allocated to this category according to the time share of it	CA, if allocated to this category according to the time share of it

317	95
374	59

$X^2=$	12.60
df=	1
p=	0.000

Table 4 Accident data (N) in the form of contingency tables, and the test results for comparing work place and home/leisure accidents.

WA, actually	HLA, actually
WA, if allocated to this category according to the time share of it	HLA, if allocated to this category according to the time share of it

317	1496
374	799

$X^2=$	83.02
df=	1
p=	0.000

Table 5 Accident data (N) in the form of contingency tables and the test results, for comparing home/leisure and commuting accidents (p=0.379)

HLA, actually	CA, actually	1496	95
HLA, if allocated to this category according to the time share of it	CA, if allocated to this category according to the time share of it	799	59

$X^2=$	0.78
df=	1

metal processing company, accident risk is statistically significantly different during time at the work place, i.e. lower, compared to commuting or home and leisure. The last mentioned two categories appear to be quite equal as far as risk (Table 5).

Accident rates among staff, production workers and supervisors (including experts), could be compared in the following way, based on the RR statistic:

$$RR = \frac{P1 : (P1 + P3)}{P2 : (P2 + P4)}$$

For WA, the proportions of Pi were the following ($\sum Pi = 1$):

	Involved in WA incl. CA	
	Yes	No
Prod. workers	$P1$	$P3$
Supervisors	$P2$	$P4$

$$RR\ (WA) = \frac{0.043 : (0.043 + 0.657)}{0.002 : (0.002 + 0.298)}$$

$$RR = 9.2$$

The estimated RR statistic (metal processing, year 2002) shows that the incidence of injurious work place accidents (including commuting) was clearly higher in blue-collar employees than in white-collar employees. The amount of CA was so minor in supervisors; there was no reliable base to estimate RR separately for them. Hence, the WA and CA were analysed together. They belong to the common accident insurance system in Finland in the same way.

As far as HLA, the proportions of *Pi* were the following ($\sum Pi = 1$):

	Involved in HLA	
	Yes	No
Prod. workers	*P*1	*P*3
Supervisors	*P*2	*P*4

$$\text{RR (HLA)} = \frac{0.104 : (0.104 + 0.596)}{0.012 : (0.012 + 0.288)}$$

$$\text{RR} = 3.7$$

The estimated RR statistic (metal processing, year 2002) shows that the incidence of injurious HLA was relatively much higher in production workers (blue-collar) than in supervisors (white-collar).

5 Discussion and Conclusions

Much research has been done to reduce accidents at the workplace and in traffic, but home and leisure-time accidents have not received as much attention as they should. In Finland, according to the current general survey ($N = 7,193$) that is conducted every 3 years, at the national level approximately 70 % of all accidents involving adults take place at home and during leisure time. Our study reveals new detailed knowledge about a corresponding proportion of the population ($N = 13,000$) employed at one large company and public sector workplaces. As far as the studied cases (metal processing and municipality) are concerned, HLA ($N = 7,464$) appeared to be the most numerous (the order of magnitude being minimum approximately 50 %) compared with accidents ($N = 4,322$) in other categories of employed people. The large share of workplace accidents was out of proportion in the municipality, and correspondingly at the industrial mill, the high incidence of HLA was out of proportion with what could be anticipated.

In the metal processing case, when considering more than 10 years of data, it can be seen that the number of WA decreased to one third, while HLA decreased only minutely, even though there could have potentially been more (number of HLA compared to WA). There is not a big difference when considering the number of CA. In the municipality case, the number of accidents increased in every category. On average, the absence days per case remained almost the same.

The average number of CA in Finland is normally around 16 % of WA (The Federation of Accident Insurance Institutions 2007). Metal processing was quite near this national average for Finnish working life. In the municipality case this percentage was significantly higher.

If all minor injuries are taken into account, the accident incidence rate was 204 in the Finnish adult population. The national survey gives figures that in many ways support the workplace-specific data. The survey gives a picture on a broader basis —all adult Finns—compared with the specific cases. When safety management needs are considered, workplace-specific "own" data is necessary.

Knowledge about HLA and CA needs to be improved among people and organizations. Employers have traditionally been interested only in accidents that take place at work, but the significance of HLA and CA is becoming incrementally clearer. Most people belong to a work community; therefore the workplace could be a good place to also inform about HLA and CA, not only AW. An important factor is to improve compilation of statistics and get more detailed information about off-the-job accidents. Thereby, risk prevention and management activities at workplaces can cover home and leisure time better, as well as CA, which are very significant as important sources of absence and non-productivity. Both cases from different work sectors showed the importance of off-the-job injuries.

Hokkeri (2011) created a workplace travel plan for Turku University Hospital. This kind of hospital is a workplace owned by a municipality in Finland. The aim was to increase the use of sustainable travel modes considering cost perspective. About 65 % of all work place travel kilometres were driven by private car; it created the highest emissions in every emission grades and was cheaper than public transportation for short distances. By encouraging employees to commute by bike or foot, an employer can also refresh its image and profile itself as an organization that is interested in employee health.

Though there are many benefits, such as healthier employees, less sick leaves and costs (for both employer and public healthcare), when commuting by, e.g. bike, it should be remembered that serious accidents take place when biking. One survey (Bench 2012) has revealed that the most common accident mechanisms are bicycling, horseback riding and team ball sports, with bicycling causing serious injury most frequently. Another research (Airaksinen et al. 2014) reveals that cycling under the influence of alcohol increases the risk of head injuries. One third of the injured cyclists were under the influence of alcohol at the time of hospital attendance. There were also more head injuries in alcohol-related accidents (60 %) than in those where the cyclist was sober.

Use of a cycling helmet—which is not obligatory or sanctioned in Finland, unlike some other countries—is not yet common. Conversely, use of safety protectors is obligatory and sanctioned at workplaces.

Reason (1997) made conclusion by presenting three approaches to safety assurance. These models for managing safety comprise the engineering model (T, technology), the organizational model (O) and the person model (P) (c.f. TOP) (Kjellen 2000; Luczak 1998; Reason 1997).

Traditional emphasis on training of persons can be characterised with direct quotations by Reason (1997) 'The person model is exemplified by the traditional occupational safety approach. The main emphases are upon individual unsafe acts and personal injury accidents. It views people as free agents capable of choosing between safe and unsafe behaviour'. In the person model: 'The most widely used countermeasures are "fear appeal" poster campaigns, rewards and punishments, unsafe act auditing, writing another procedure, training and selection. Progress is measured by personal injury statistics such as fatalities, lost-time injuries, medical treatment cases, first aid cases, and the like. It is frequently underpinned by the "iceberg" or "pyramid" views of accident causation'. On the contrary, when (O) is emphasised, i.e. the company level, the community has to be seen as a very important systemic context: 'The organizational model views human error more as a consequence than as a cause. Errors are symptoms that reveal the presence of latent conditions in the system at large. They are important only in so far as they adversely affect the integrity of the defenses'. The P approach to control, in addition to T, is available as far as all accident categories are concerned. The important O approach is available in its full effect only in WA. Meanwhile, in a weaker way, it is also usable in CA. But, when a typical HLA is thought, the controlling force of 'O means' is quite minimal, at least nowadays, when HLA are not known and discussed at the workplace.

The holistic reality of accident risks has to be brought to people's attention, and personal awakening is needed. At the level of each individual the total risk of all accidents is a key factor of well-being and ability to work. Prevention methods can be better focused on when the causes of accidents and absences are known. One's own proactive approach safety at the workplace, while commuting and at home, is important. Everyone should be involved.

Hollnagel (2014) represents a new model for accident prevention (Safety-II). Traditionally, when studying accidents and their causes, concentration has centred on 'why' something happened. The basic idea has been that there are different reasons for failures and success. Hollnagel calls this as Safety-I: safety has been defined by using its opposite, lack of safety. Safety-II concentrates on situations when everything has gone well, and 'safety has been involved'. In the Safety-II-model, unlucky and lucky situations have the same reasons: difference in action. In Safety-II, safety means the ability to manage in varying conditions and adjust to circumstances, to manage activities toward the best possible results.

Also Harms-Ringdahl (2013) represents a safety model based on energy: 'Energy Analysis is based on a simple idea: for an injury to occur, a person must be exposed to an injurious influence—a form of energy'. Energy can be understood as something that can damage a person (e.g. physically or chemically) in connection with a particular event. The concept of barriers is also one essential part of the Harms-Ringdahl energy model. Barriers prevent energy from coming into contact with the person and/or causing injury. According Harms-Ringdahl in the energy model, an injury occurs when a person or object comes into contact with a harmful energy, and this means that the barriers have not provided sufficient protection. Could these models also bring about new ideas for home and leisure safety?

More comprehensive and nationwide studies on all accidents are clearly needed, at least in Finland. Therefore, support can be found for the recommendation that 'safety-conscious' workplaces should also include home and leisure-time accidents in their risk prevention and safety management programmes, as some companies say they do (Mottel et al. 1995). Assessment and auditing systems should also more frequently include home and leisure accident records like the International Safety Rating System ISRS (Kjellen 2000). Computerised SHE, i.e. Safety, Health and Environmental information systems within organizations (Kjellen 2000; Räsänen 2007) should include HLA, as well as CA. HLA and CA are quite well included in the safety management and information systems of the metal processing mill case. Even if HLA recordings are done, many important features linked to modern WA accident prevention are difficult or maybe impossible to carry out as far as HLA or CA are concerned: they include near-accident or hazardous situation follow-ups and risk assessments. These have often been very effective tools, and so are risk analyses (Harms-Ringdahl 2001; Kjellen 2000), which are difficult to implement to cover HLA. Total holistic safety management and information should be gradually integrated into general daily management systems. One possibility is to integrate them into Enterprise Resource Planning (ERP) systems.

At the national level it might be useful to utilise EuroSafe much more in the future, which also records information about accidents in private life (EuroSafe 2006). EuroSafe, the European Association for Injury Prevention and Safety Promotion, is a network of injury prevention champions dedicated to making Europe a safer place. Occupational health care systems in workplaces could have the proper expertise to utilise the systems provided by EuroSafe. Though there currently exists a national web portal of HLA in Finland (Tapaturmaportaali 2008), the significance of HLA, especially at work places, needs a new type of statistics and effective media to communicate the real, total accident figures. This would support holistic and consistent safety culture.

It has been said that finding more efficient ways of preventing accidents and injuries is a real challenge (Harms-Ringdahl 2013). On the other hand, HLA are not often even mentioned in keyword lists of safety books, despite a few exceptions (Mottel et al. 1995; Harms-Ringdahl 2013). In general, home and leisure safety should be discussed in safety books.

An accident and injury recording system has been established in the USA (National Safety Council 2006a, b). There, as in Finland, two out of three accidents among working-aged citizens occur during leisure time (National Safety Council 2006b). Costs and other losses linked with all accidents are significant as far as the individual employee; society and the workplace are concerned. In many cases, the number of home and leisure-time accidents has increased, while accidents at work have decreased at large companies. Could the risk compensation theory give an explanation for that? Safety thinking should be 'present' 24 h a day, 7 days a week. To achieve an overall safety culture, more awakening at societal and workplace levels is needed. In addition, each individual should have a special interest in comprehensive accident risk as a potential factor that threatens his/her well-being and quality of life.

References

Airaksinen N, Nurmi-Lüthje I, Lüthje P (2014) Pyöräily alkoholin vaikutuksen alaisena lisää pään vamman riskiä (Cycling under the influence of alcohol increases the risk of head injuries). Suomen lääkärilehti 18/2014, pp 1313–1318

Bench F (2012) Multidetector computed tomography of spinal and pelvic fracture with special reference to polytrauma patient. Academic dissertation, University of Helsinki

EuroSafe (2006) European association for injury prevention and safety promotion. http://www.eurosafe.eu.com/csi/eurosafe2006.nsf. Accessed 3 March 2008

Haikonen K, Lounamaa A (eds) (2010) Suomalaiset tapaturmien uhreina 2009, kansallisen uhritutkimuksen tuloksia. Terveyden ja hyvinvoinnin laitos (THL), Raportti 13/2010

Harms-Ringdahl L (2001) Safety analysis. Principles and practice in occupational safety. CRC Press, Boca Raton

Harms-Ringdahl L (2013) Guide to safety analysis for accident prevention. IRS Riskhantering AB, Stockholm

Hokkeri S (2011) Työmatkaliikkumissuunnitelma Turun yliopistolliselle Keskussairaalalle. Master's thesis, Lappeenranta University of Technology

Hollnagel E (2014) Nollatoleranssista resilienssiin? Suomalainen yhteiskunta, muutos ja turvallisuus - seminaari 12.3.2014. http://www.spek.fi/Suomeksi/Ajankohtaista/Tutkimustoiminta/Resilienssi. Accessed 20 April 2014

Kansallinen uhritutkimus 2006 (National Victim Survey 2006) (2007) Helsinki:Kansanterveyslaitos (National Public Health Institute in Finland). http://www.ktl.fi/portal/suomi/osastot/eteo/yksikot/koti_ja_vapaa-ajan_tapaturmien_ehkaisyn_yksikko/uhri2006/. Accessed 3 March 2008

Kjellen U (2000) Prevention of accidents through Experience feedback. Taylor & Francis, London

Luczak L (1998) Arbeitswissenschaft. Springer, zweite Auflage, Berlin

Ministry of Social Affairs and Health (2005) Research on well-being at work in Finland—focusing on occupational health and safety. Report 25. Ministry of Social Affairs and Health, Helsinki

Mottel WJ, Long JF, Morrison DE (1995) Industrial safety is good business. The Dupont story. Wiley, New York

National Safety Council (2006a) Injury Facts 2005–2006 edition. National Safety Council, Itasca

National Safety Council (2006b) The off-the-job safety program manual, 2nd edn. National Safety Council, Itasca

Qvist E, Väyrynen S (2005) Workplace, commuting, and home & leisure-time accidents among employees of a steel mill. In: Veiersted B, Fostervold KI, Gould KS (eds) Proceedings of NES 2005. Ergonomics as a tool in future development and value creation. 37th annual congress of the Nordic ergonomics society NES, 10–12 Oct 2005. NES, Oslo, pp 337–340

Qvist E, Väyrynen S, Härönoja R (2004) Terästehtaan henkilöstön työ- sekä koti- ja vapaa-ajan tapaturmat (Work, home and leisure accidents among personnel of steel works). Työ & Ihminen (18)2:257–268

Räsänen T (2007) Management of occupational safety and health information in Finnish production companies. Finnish Institute of Occupational Health, Helsinki

Reason J (1997) Managing the risks of organisational accidents. Ashgate Publishing Limited, Aldershot

Sommer H, Brügger O, Lieb C, Niemann S (2007) Volkswirtschaftliche Kosten der Nichtberufsunfälle in der Schweiz. Strassenverkehr, Sport, Haus und Freizeit. Bfu-report 58

Tapaturmaportaali (Accident Portal) (2008) Kansanterveyslaitos (National Public Health Institute in Finland), Helsinki. http://www.ktl.fi/portal/suomi/muuta/oikopolut/tapaturmaportaali/. Accessed 3 March 2008

The Federation of Accident Insurance Institutions (FAII) (2007) Työtapaturmat ja ammattitaudit – Tilastovuodet 1996–2006 (Occupational accidents and diseases in Finland 2006—review of trends in the statistical years 1996–2004). The Federation of Accident Insurance Institutions, TVL, Helsinki

The Statistics Finland (2007) Työtapaturmatilasto 1996–2005 (Occupational accidents in Finland 1996–2005). http://www.stat.fi. Accessed 5 May 2007

Väyrynen S, Hoikkala S, Ketola L, Latva-Ranta J (2008) Finnish occupational safety card system: special training intervention and its preliminary effects. Int J Technol Human Interact (4) 1:15–34

Yrjämä L, Ollanketo A (eds) (2007) Henkilöstön koti- ja vapaa-ajan tapaturmien ja niiden torjunnan merkitys työyhteisöissä: tapausesimerkeistä kohti yleistävää tietoa, KOTVA–hanke (Work, home and leisure accidents among personnel of some big work places). Project report 25. Oulu University Press, Oulu

Yrjämä L, Väyrynen S (2006) Home & leisure time accidents compared with work accidents among employees of two paper mills. In: Saarela KL, Nygård C-H, Lusa S (eds) Proceedings of NES 2006. Promotion of well-being in modern society. 38th annual conference of the Nordic ergonomics society NES, 24–27 Sept 2006, Hämeenlinna, Finland. University of Tampere, Tampere, pp 241–244

Chapter 10
HSEQ Training Park in Northern Finland—A Novel Innovation and Forum for Cooperation in the Construction Industry

Arto Reiman, Olli Airaksinen, Seppo Väyrynen and Markku Aaltonen

Abstract Safety training in the construction industry needs new methods and procedures if it is to effectively reduce currently high accident statistics figures. The HSEQ Training Park is a novel safety training innovation that enables practical demonstrations and active participation. The Park was constructed in Northern Finland through the cooperation of almost 70 companies and communities. It began its activities with trainer trainings in the spring of 2014. This study provides an overview of the design, construction process and structure of the Training Park.

Keywords Accidents at work · Construction industry · HSEQ Training Park · Participation · Safety at work · Safety innovation · Training

---

A. Reiman (✉)
Finnish Institute of Occupational Health, Aapistie 1, 90220 Oulu, Finland
e-mail: arto.reiman@ttl.fi

O. Airaksinen
HSEQ Park Northern Finland, Hiltusentie 9, 90550 Oulu, Finland
e-mail: olairaks@gmail.com

S. Väyrynen
Faculty of Technology, Industrial Engineering and Management, University of Oulu,
P.O. Box 4610, 90014 Oulu, Finland
e-mail: Seppo.Vayrynen@oulu.fi

M. Aaltonen
Finnish Institute of Occupational Health, Arinatie 3A, 00370 Helsinki, Finland
e-mail: markku.aaltonen@ttl.fi

1 Introduction

The construction industry covers about 7 % of the employed labour in Finland (in 2012, 175 000 employees), and is male-dominated, as approximately 90 % of the workforce are men (Oksa et al. 2013). According to the statistics of the Federation of Accident Insurance Institutions (FAII), the construction industry has the second highest accident figures in the comparison between the absolute accident numbers of different branches. The amount of accidents at work has remained quite stable in recent years, at approximately 14 000 accidents/year. Accident frequency is over 60 accidents/million working hours.

During recent years, the largest construction companies have succeeded in lowering their accident figures, whereas accident statistics remain especially high in the construction industry's SMEs (FAII 2013; Confederation of Finnish Construction Industries RT 2014). In addition to high accident figures, construction work contains several physical load factors, such as difficult and repetitive working positions, manual lifting and transfers, and whole-body and hand-arm vibrations, which may cause work-related musculoskeletal disorders (Albers and Estill 2007; Boschman et al. 2012; Rwamamara et al. 2010).

In their review, Pinto et al. (2011) listed several causes for poor safety performance in the construction industry. These include poor safety organization, lack of coordination at shared workplaces, tight schedules and economic pressure, constantly changing work environments, specialization in certain tasks, poor safety awareness among top management and project managers, poor involvement of workers in safety matters, lack of and/or inadequate prevention/protection equipment, and lack of proper communication procedures. They also point out that company size seems to affect safety performance, as smaller companies tend to have limited resources for health and safety issues (2011). In Nordic countries, additional threats, such as changing environmental conditions, slipperiness, darkness and cold must all be seen as additional risk factors (McFadden and Bennett 1991; Risikko 2009).

Guo et al. (2012) emphasize inadequate safety training as one of the main factors behind poor accident statistics. Glendon et al. (2006) also claim that safety training is a great deal more efficient if pictures and demonstrations are utilized rather than mere lectures or reading of the relevant material. Active participation and simulation can also boost the efficiency of training.

Construction work often takes place at shared workplaces, where different actors work together (see e.g. Väyrynen et al. 2008). Thus, development work requires broad participation and commitment and in-depth understanding of the causes of all stakeholders' accidents (Gervais 2003; Sousa et al. 2014). Traditional technical protection solutions, rules and supervision are often seen as the main methods for safety management (Törner and Pousette 2009), but safety at work can be improved by various other means. A novel approach to safety training has been introduced in Finland. Two HSEQ (Health, Safety, Environment, Quality) Training Parks have been constructed by stakeholders in the construction industry. The first Training

Park of this kind in Europe, namely Rudus Safety Park, was constructed by multinational construction company Rudus (RUDUS Safety park 2014) in the city of Espoo in Southern Finland in 2009. This was followed by the HSEQ Training Park in Northern Finland in the city of Ouluin 2013 (PSTP 2014; YLE 2014). The Oulu Park has not yet been officially named in English yet. In this article, the Oulu Park is discussed under a working title HSEQ Training Park in Northern Finland.

The HSEQ Training Park concept is a unique safety innovation by which different actors of the construction industry and other branches can be trained on a practical level to perform different work phases safely at construction sites. Both Training Parks comprise different *training points*, which contain different practical, real working life demonstrations. The most notable difference between the two HSEQ Training Parks is that the Oulu Park was constructed jointly by several different stakeholders, whereas the Training Park in Espoo was originally constructed by the Rudus company. About 5 000 workers are trained in the Espoo Training Park each year. The Oulu Training Park aims for a similar level of annual attendance (Fig. 1).

Almost 70 companies and communities were involved in the joint effort of constructing the HSEQ Training Park in Oulu. The construction industry is strongly involved in the development of the Training Park. Other industry and service branches have also played a big part in the Training Park design and construction process. A full, updated list of the participants involved is available on the Training Park website (PSTP 2014). The Park is governed by the Turvapuisto (Training Park) association of Northern Finland.

The main objectives of this study are as follows:

1. To present the design and construction process for the Training Park in Northern Finland,
2. To introduce results based on trainees' assessments of the contents and realizations of the training points in the Training Parkin Oulu.

Fig. 1 An overview of the Training Park in Oulu. In this picture (in the *left side*), a group of workers is being trained at the construction work for foundations and framework training point

2 Study Design and Methods

2.1 Training Park Design and Construction Process

The Training Park design and construction process was carried out in 2013−2014. Each training point was designed and constructed by the companies and communities, and has a master organization supplemented by other voluntary organizations. The whole Training Park design and construction process was controlled by a project manager.

2.2 Questionnaire

A questionnaire was developed for the purposes of the study, which was aimed at those participating in the first three trainer training sessions in the Oulu Training Park. Three trainers and around 20 trainees attended each training session. The questionnaire is based on the design science framework (Van Aken 2004), as it provides innovation evaluation knowledge for further development processes concerning the Training Park. The purpose of the questionnaire was to gather (1) quantitative assessments on the content and realization of each particular training point in the Training Park, and (2) qualitative initiatives for future development issues in the Training Park. The quantitative assessments were made on a scale from 1 (a great deal to improve) to 5 (well-constructed). The questionnaire was filled in on paper by the trainees during the training session.

Altogether 62 trainees responded to the questionnaire. Twelve questionnaires were inadequately filled and were thus rejected. The majority of the respondents (44; 88 %) were male. One-third (17; 34 %) of the respondents belonged to top management, administration or work supervision (12; 24 %), three (6 %) of the respondents represented employees, and the rest (18; 36 %) were different kinds of trainers and educators.

3 Results

3.1 Training Park Design and Construction Process

The participatory design process of the HSEQ Training Park has led to 19 different training points to date. The training point topics and companies and communities that have been involved in the construction of the particular training point are shown in Table 1. The numbering of the training points is based on the layout of the

Table 1 Training points, companies and communities (in alphabetical order) that participated in the construction process

Training point	ID number on layout	Companies and communities involved in training point design and construction
Safety management and planning	A-1	Confederation of Finnish Construction Industries RT; Finnish Construction Trade Union; Infra union; Finnish Institute of Occupational Health; Pohto Institute; Regional State Administrative Agencies; Skanska Rakennuskone Inc; University of Oulu
Access control and grey economy	A-2	Construction Quality Association RALA; Cramo Inc; Suomen tilaajavastuu Inc; Trade Union PRO
Safety exhibition for personal protective equipment	1	ETRA Inc; If P&C Insurance; Lujatalo Inc; Sievi Inc; Würth Inc
Construction work for foundations and framework	2	Lujatalo Inc; Rajaville Inc; Ramirent Inc; Ruskon Betoni Inc; Skanska Talonrakennus Inc
Construction work for house technology	3	Lemminkäinen talo Inc; Lemminkäinen talotekniikka Inc
Tools, working levels and personal passenger hoists	5	Lujatalo Inc; Ramirent Inc
Construction work inside houses	6	Meranti Inc; NCC group; Ramirent Inc; Rudus Inc
Reconstruction work	8	Temotek Inc; YIT Rakennus Inc; Telinekataja Inc; Paupek Inc
Property maintenance	9	City of Oulu; Icopal Inc
Transits and transportations at sites	11	Rudus Inc; Kuljetus Polar Inc
Excavation protection	12	Skanska Infra Inc
Dangers in excavation work	14	Oulun Energia; Oulun Vesi
Traffic control in road work	15	Lemminkäinen infra Inc
Dangers of overhead lines	16	Fingrid Inc; Oulun Energia
Asphalt work	18	Skanska Asfaltti Inc
Fire safety training	26	Temrex Inc
Construction within industrial processes	28	Ruukki Metals plc; Miilukangas Inc
Lifting and hoisting safety	29	Havator Inc; Hartela Inc; SRV Inc
Single-family house construction work	30	Building inspection authority; DesignTalo; Regional State Administrative Agencies

Training Park. Thus, some numbers (such as 4, 7, 10, etc.) have not yet been taken into use.

Each training point constitutes a separate training environment, and has several detailed topics. Examples of the training points are shown in Figs. 2 and 3.

Fig. 2 Training points. Incorrect working methods and inadequate personal protection for chipping old lavatory wall (Training point 8, *left*) and the casting of a floor (Training point 6, *right*). The *dark clothed* manikins represent incorrect working methods and inadequate personal protection, whereas the other two manikins represent correct working methods and equipment

Fig. 3 Specific details at training points. Good practices related to floor casting and floor supports (Training point 2, *left*) and proper scaffolding (Training point 30, *right*)

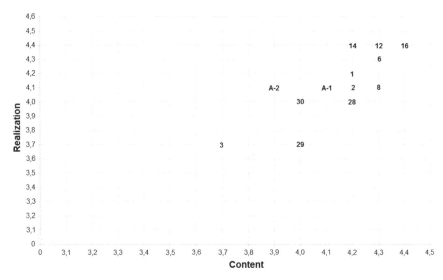

Fig. 4 Trainees' assessments of the content and realization of each training point that was ready for the trainer training sessions. Quantitative assessments were made on a scale of 1 (a great deal to improve) to 5 (well-constructed). The numbering in the figure responds to the numbering in Table 1

3.2 Questionnaire

The assessment of the content and realization of each training point in the Training Park are shown in Fig. 4. Training points 5, 9, 11, 15, 18, and 26 (see Table 1) were not ready for the trainer training sessions, and were thus excluded from the assessments.

Each of the assessed training points was given values below 3.5 for both their realization and content (Fig. 4). Training point 16: *Dangers of overhead lines* was given the highest number for both content and realization in the trainee assessments. Training points 6: *Construction work inside buildings*, 12: *Excavation protection* and 14: *Dangers in excavation work* were also given high assessment values for both content and realization.

4 Discussion and Conclusions

If occupational safety in the construction industry is to be improved, new innovative concepts for safety management and training are needed. Construction work is done in work environments in which employees cannot be continuously supervised. Thus in many cases, employers must place their trust in their employees' ability to perform work safely in all circumstances. Holistic, systemic safety training is one way in which to enhance employees' abilities and knowledge regarding this topic.

The concept of the HSEQ Training Park as a new novel safety training innovation has been introduced in Finland. The original concept by the Rudus Construction Company in Espoo in Southern Finland was followed by a joint development process in Northern Finland, in Oulu. The construction process of the Training Park in Oulu shows how rival companies can jointly develop new kinds of practices when all stakeholders have a common interest in accident-free construction sites. A good benchmark for the construction industry as regards cooperation could be the Northern Finnish process industry, which has for years successfully jointly developed HSEQ practices, as reported by Väyrynen et al. (2012).

Around 20 training points are in use currently and a few more training points are under construction and will be finalized during the year of 2014. The Oulu Training Park is now in use. Over 100 trainers have been trained to teach HSEQ issues at the training points, and common training material has been created to provide uniform training processes. Several educational institutions and organizations in Northern Finland have adopted Training Park training into their curriculums.

Numerous new ideas regarding future needs for training practices and development activities have been raised, for example, during the Training Park construction phase and in the trainer training sessions, and in organizations' and communities' own training sessions. These thoughts include a willingness to ensure that the construction industry's SMEs also apply Training Park training in their safety management practices. New practices for these purposes are planned for execution in the next few years.

This study presents results from the questionnaire distributed in the trainer training sessions. The questionnaire served as a preliminary tool for evaluating the Training Park innovation. However, the results should be examined with caution, since the teaching style of the three trainers who provided the trainer trainee sessions might have varied. Furthermore, some parts of the training points that were not quite ready may have affected the assessments. The organizations responsible for the training points have been informed about the questionnaire results and are now using them to improve the Training Park. A demanding new research challenge in the Training Park is to study how effective the training actually is. New plans for effectiveness studies are under consideration for implementation in the next few years.

Acknowledgments We wish to express our gratitude to the partners of the HSEQ Training Park Northern Finland association. Furthermore, we would like to thank the trainees who participated in the training point assessments.

References

Albers JT, Estill CF (2007) Simple solutions. Ergonomics for construction workers. DHHS (NIOSH) publication no 2007-122

Boschman JS, van der Molen HF, Sluiter Jk, Frings-Dresen MHW (2012) Musculoskeletal disorders among construction workers: a one-year follow-up study. BMC Musculoskel Dis 13 (196):1471–2474

Confederation of Finnish Construction Industries RT (2014) Bulletin in Finnish: Rakennusteollisuuden ensimmäinen yhteinen työturvallisuusviikko: lupaavia tuloksia turvallisuuteen satsanneissa yrityksissä. http://www.rakennusteollisuus.fi/Ajankohtaista/Tiedotteet1/2014/Rakennusteollisuuden-ensimmainen-yhteinen-tyoturvallisuusviikko-lupaavia-tuloksia-turvallisuuteen-satsanneissa-yrityksissa/. Accessed 4 June 2014

FAII (2013) Tilastojulkaisu. http://www.tvl.fi/fi/Tilastot/Tilastojulkaisut/Tilastojulkaisu/. Accessed 3 June 2014

Gervais M (2003) Good management practice as a means of preventing back disorders in the construction sector. Saf Sci 41(1):77–88

Glendon AI, Clarke S, McKenna EF (2006) Human safety and risk management. CRC Press, Taylor & Francis, Boca Raton

Guo H, Li H, Chan G, Skitmore M (2012) Using game technologies to improve the safety of construction plant operations. Accid Anal Prev 48:204–213

McFadden TT, Bennett FL (1991) Construction in cold regions. A guide for planners, engineers, contractors and managers. Wiley, New York

Oksa P, Savinainen M, Lappalainen J (2013) Rakentaminen. In: Kauppinen T, Mattila-Holappa P, Perkiö-Mäkelä M, Saalo A, Toikkanen J, Tuomivaara S, Uuksulainen S, Viluksela, M, Virtanen S (eds) Työ ja terveys Suomessa 2012. Seurantatietoa työoloista ja työhyvinvoinnista. Finnish Institute of Occupational Health, Helsinki, pp. 183–186

Pinto A, Nunes IL, Ribeiro RA (2011) Occupational risk assessment in construction industry—overview and reflection. Saf Sci 49(5):616–624

PSTP (2014) HSEQ training park in Northern Finland web site. http://www.pohjois-suomenturvapuisto.fi/. Accessed 3 June 2014

Risikko T (2009) Safety, health and productivity of cold work. A management model, implementation and effects. Dissertation, University of Oulu

RUDUS Safety park (2014) Company web page. http://www.rudus.fi/turvapuisto. Accessed 3 June 2014

Rwamamara RA, Lagerqvist O, Olofsson T, Johanssom BM, Kaminskas KA (2010) Evidence-based prevention of work-related musculoskeletal injuries in construction industry. J Civ Eng Manag 16(4):499–509

Sousa V, Almeida NM, Dias LA (2014) Risk-based management of occupational safety and health in the construction industry—Part 1: Background knowledge. Saf Sci 66(1):75–86

Törner M, Pousette A (2009) Safety in construction—a comprehensive description of the characteristics of high safety standards in construction work, from the combined perspective of supervisors and experienced workers. J Saf Res 40(6):399–409

Van Aken JE (2004) Management research based on the paradigm of the design sciences: the quest for field-tested and grounded technological rules. J Manage Stud 41(2):219–246

Väyrynen S, Hoikkala S, Ketola L, Latva-Ranta J (2008) Finnish occupational safety card system: special training intervention and its preliminary effects. Int J Tech Hum Interact 4(1):15–34

Väyrynen S, Koivupalo M, Latva-Ranta J (2012) A 15-year development path of actions towards an integrated management system: description, evaluation and safety effects within the process industry network in Finland. Int J Strat Eng Asset Manag 1(1):3–32

YLE (2014) Finland's national public service broadcasting company. http://yle.fi/uutiset/oulun_turvapuisto_aloitti_raksatyon_turvakoulutuksen/7153864. Accessed 3 June 2014

Part III
Effects of the OSHM

Chapter 11
Safe Use of Chemicals and Risk Prevention in the Finnish Chemical Industry's Work Places

Toivo Niskanen

Abstract The aim of the present study was to examine how HSE (health, safety and environment) managers and the workers' OSH representatives view the following topics: (1) safe use of chemicals, (2) prevention prioritizations and measures of management, (3) collective climates of OSH in systems thinking and (4) discourses on legislation about chemical safety. It was hoped to clarify their potential effects by evaluating organizational relationships and technical measures of OSH. Forty-nine HSE managers and 105 workers' OSH representatives responded to an online survey questionnaire. In the regression analysis of HSE managers and workers' OSH representatives, both the prevention prioritizations and the measures of management displayed a highly significant ($p < 0.001$) effect on safe use of chemicals. With respect to safe use of chemicals, the following factors were found: Factor 1—Training, Factor 2—OSH Policies and Factor 3—Work processes. With respect to prevention prioritizations, the following factors were found: Factor 1—Technology, Factor 2—Assessments and documents and Factor 3—Control. For measures of management, the factors were: Factor 1—Leadership, Factor 2—Assessments and Factor 3—Utilization of measures. Qualitative results in the discourses detected an improvement with respect to compliance with OSH legal demands, need of information and cooperation between different governmental inspectorates.

Keywords Systems · Climate · Risk · Assessment · Prevention · Safety · Health · Chemical industry

T. Niskanen (✉)
Ministry of Social Affairs and Health, Occupational Safety and Health Department, Legal Unit, P.O. Box 33, FI-00023 Government, Helsinki, Finland
e-mail: toivo.niskanen@stm.fi

1 Introduction

1.1 Risk Assessments and Prevention Measures

Van Asselt and van Bree (2011) stated that in "risk governance" it is important to develop a framework and tools and to design processes, guidelines and institutions that enable society to adequately deal with modern risk; this represents an ongoing challenge to the risk community. Technological advances in the form of a new product or industrial process often pose risks to health, safety and the environment (Baram 2007; Baker et al. 2007). Process control and safeguarding equipment has become more complex, allowing flexibility and overview at a higher level but related a corresponding increasing risk of faulty use of chemical agents (Knegtering and Pasman 2009).

The precautionary principle should affect all stages and all aspects of dealing with risk, from risk framing, to risk assessment, communication, management and regulation (van Asselt and van Bree 2011). This is intended to confer freedom on companies working in areas with responsibilities—but also the responsibility—to decide on the management of their operations as long as they act within broadly defined safety functions and goals. Instead of having to continuously react to changes in external demands and adapt their safety management accordingly, companies can manage safety in a proactive manner in a way that best fits their specific situation (Grote 2012).

Grote (2012) examined high-risk industry and found that three attributes of organizations and their environments are very directly related to how safety can and should be managed: (1) The kinds of safety to be managed: Process versus personal safety; (2) The general approach to managing uncertainty as a hallmark of organizations that manage safety: Minimizing versus coping with uncertainty; and (3) The regulatory regime within which safety is managed: External regulation versus self-regulation. Consequently, hazards associated with work activities can be present as a result of any one or a combination of the following: substances, machinery/processes, work organization, tasks, procedures, the employees and circumstances in which the activities take place (Gadd et al. 2004). However, instead of allocating all resources to generate information on present hazards, resources should also be available to generate information related to safer technology alternatives (Helland 2009).

Risk assessment should be an integral part of the company's safety management system. An assessment will almost inevitably result in recommendations for improvements and further to be taken actions to control and reduce risk. Furthermore, any identified new or additional risk reduction measures or risk control systems must be implemented. When safety is controlled by numerical performance objectives as is the case with generic regulation, safety becomes just another criterion of a multi-criteria decision making and it becomes an integral part of normal operational decision making within the line organizations (Rasmussen 1997). It is important that the organizations should have a coherent and systematic way of considering possible safety problems and of ways to avoid the risks (Hale et al. 2007).

Pasman et al. (2009) concluded that the essence of process safety is to be aware of hazards, to estimate the risks, to reduce risks where possible, identify the signals when danger is becoming imminent, and to know what to do to neutralize any threats. For example, recommendations (Baker et al. 2007) can be based on an ethical approach; according to this report the legal/regulatory process is not sufficient, it must be supported by ethical and cultural values. It is evident that safety control requirements and a new approach to representation of system behaviour is necessary focussed on the mechanisms generating behaviour in the actual, dynamic work context (Rasmussen 1997). Mostia (2009) concluded that a risk reduction strategy should have the following core attributes: (1) A holistic approach to risk; (2) Responsibilities, accountabilities and authority defined; (3) Continuous improvement program; (4) Risk reduction as part of the normal work process; and (5) Risk reduction as a core attribute of the safety climate/culture.

Goal rules only define the goal to be achieved, leaving open how this is accomplished by the actors concerned (Grote 2012). Process rules provide guidance for deciding on the right course of action for achieving certain goals. Finally, action rules prescribe detailed courses of action, possibly without even mentioning the goal to be achieved. As a rule of thumb on good rule making, action rules should be used when stability of processes is required. Goal and process rules should be used when flexibility is required. Communication can be crucial in sharing of information about the risks and possible ways of handling them. It might support building and sustaining trust among various actors through which particular arrangements or risk management measures become acceptable. It might result in actually involving people in risk-related decisions, through which they gain ownership. The integration principle calls attention to the need to consider the interconnections, both content-wise and in terms of process, between the various risk-related activities (van Asselt and Renn 2011).

1.2 OSH Climate

Mearns et al. (2010) concluded that investment by an organization in OSH policy beyond mandatory requirements might well be reciprocated by workers through compliance with safety rules that will benefit not only themselves but also the workplaces as a whole. According to Rochlin (1999), organizations with exemplary safety performance, especially those in high hazard industries (e.g. high-reliability organizations), display a positive engagement with safety that extends beyond conventional safety technologies and controls and in fact represents an active process of anticipating and planning for unfortunate events and circumstance. When the focus is on self-regulation rules, companies have to find out for themselves to a large extent what kind of safety management best fits their particular situation and be more proactive in monitoring their safety performance. Compliance is determined not by adherence to detailed prescriptions but by whether companies have acceptable plans for achieving good safety performance and by actually reaching set safety goals.

Zohar and Luria (2005) have demonstrated that workers' attitudes are targeted at specific levels within the organizational hierarchy. Major safety attitudes have been found to relate to safety climate, which is defined as worker perceptions of policies, procedures and practices concerning safety in organizations (Zohar 1980). Safety compliance is defined as rule-following in core safety activities (Griffin and Neal 2000) and it has been demonstrated to be related to the safety climate. In essence, safety climate reflects the value of safety in the work environment (Neal et al. 2000). It is generally acknowledged that if there is agreement among workers about the value placed by top management on the safety of the work environment, then this is reflected in the safety climate, which becomes a characteristic of the worksite or organization as a whole. One type of behaviour that can have an effect on safety performance is compliance with and adherence to organizational rules, regulations and procedures, even when these are not being monitored (Podsakoff et al. 2003).

Zohar (2008) has argued that multi-level and multi-climate extensions represent an important link to the literature about high-reliability organizations, which has been generally overlooked in safety climate research. High-reliability organizations must cope with technological complexities; this kind of organizations includes e.g. chemical installations (Zohar 2008). Safety compliance involves adherence to rules and procedures, developed by the company and/or regulatory bodies (Zohar 2008) and is defined as rule-following in core safety activities and has been demonstrated to be related to safety climate (Mearns et al. 2010). It provided with an updated body of safety rules, which are effectively enforced, employees are likely to consider safety as a high-priority topic (Zohar 2008). The results of Neal et al. (2000) provided support for safety climate as a mediator of the impact of general organizational climate on safety-related outcomes.

Policies and procedures, according to the multi-level interpretation, define broad organizational goals and tactical guidelines relating to them, while practices relate to the implementation of organizational-level policies and procedures in sub-units. Zohar (2008) refers to shared perceptions of an organization's policies, procedures, practices and priorities with respect to safety. The organizational citizenship may be especially important for achieving and maintaining high levels of organizational safety performance in complex and demanding safety environments (DeJoy et al. 2010). Payne et al. (2010) concluded that in a process industry where there is not a good safety climate, workers are less likely to speak up about problems that could be corrected with better engineering, and management is less likely to invest the time and money to implement these changes.

Grote (2012) stated that the move towards goal-oriented regulation can also be understood in the context of the general trend towards acknowledging the need to cope with uncertainty instead of trying to manage it away. When the focus is on external regulation, companies have to continuously react to changes of regulation demands and adapt their safety management accordingly. In the chemical industry, with increasing complexity of socio-technical systems, anticipation and perfect prediction as basis for legal regulation becomes less feasible. Instead, resilience through fast adaptation to emerging threats is required which relies on decentralized small-scale trial and error.

2 The Aim of the Study and the Hypotheses Examined in the Study

The aim of the present study was to examine the relationships of the prevention prioritizations, safe use of chemicals, and measures of management. With respect to these variables also the collective climate issues were classified with the factor analysis. The reference frame of the study is presented in Fig. 1. As the background the review literature of the Hypotheses 1–3 are described in the Sects. 2.1–2.3.

As the background the review literature of the Hypotheses 1–3 are described in the Sects. 2.1–2.3.

2.1 Prevention Prioritizations

Schupp et al. (2006) indicated that design for safety in the chemical industry is becoming a more explicit and better organized process and i.e. it now applies existing knowledge about risk control and systematically seeks to learn from this new knowledge. Regulation is the crucial approach for reducing the risks posed by industrial processes which discharge pollutants into air and water and generate toxic wastes during their routine operation. As a result, the design and operation of many industrial processes are shaped by a multiplicity of regulations such as construction and operating permit requirements, ambient environmental standards and discharge

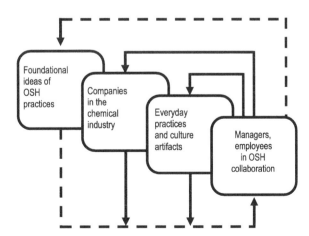

Fig. 1 The reference frame of the study

limitations and detailed rules for waste handling and disposal (Baram 2007). For example, according to Article 5 EU Directive 98/24/EC (EU Directive 1998) OSH risks shall be eliminated or reduced to a minimum by the following measures: (1) the design and organization of systems of work, (2) the provision of suitable equipment, (3) reducing to a minimum the duration and intensity of exposure and (4) reducing the quantity of dangerous chemical agents and (5) suitable working procedures.

In each control loop, at each level of the socio-technical control structure, hazardous behaviour results from inadequate enforcement of constraints on the process controlled at the level (Dulac and Leveson 2004). Contingent reward leadership means that the supervisor should be able to communicate goals and objectives to the workers, to actively monitor the worker performance towards these goals, and to provide feedback to workers (Kapp 2012). A problem-solving activity involves achieving a goal by selecting and using strategies to move from the current state to the goal state (Leveson 2000). Dulac and Leveson (2004) showed that as the complexity of engineered systems increases, hazard analysis techniques have continued to lag behind the state-of-the-art engineering practice. However, many aspects of managing risks and generating safer technologies are often seen as two different activities that are normally not aggregated in regulatory procedures or industrial risk management (Helland 2009). Another way to categorize risk is to divide the risk into two categories: identified risk (specific and general) and unidentified risk (Mostia 2009). So, this may seem like a rather obvious distinction but it does identify risk that can be dealt with directly and those risks that we have to be tackled indirectly.

Regulatory or control action involves imposing constraints upon the activity at one level of a hierarchy. Those constraints define the 'laws of behaviour' at that level that yield activity meaningful at a higher level e.g. emergent behaviour (Leveson 2000). The approaches to safety seem to put emphasis on management functions, guidelines, industry standards, quality principles, to establish the safety management system of organizations (e.g. BS 8800 2004; ILO 2001; OHSAS-18002 2000). These approaches may represent a step forward to managing safety but may not be suffice to address the management of risk in an effective manner. The IEC-standard (IEC 1995) includes technically oriented terms for a set of related terms. Risk is defined as a combination of the frequency, or probability, of occurrence and the consequence of a specified hazardous event. Risk assessment is the overall process of risk analysis and risk evaluation. Risk management and safety management are used in different ways and are often seen as identical. The IEC-standard (1995) provides the definition that risk management is the systematic application of management policies, procedures and practices to the tasks of analysing, evaluating and controlling risk.

2.2 Safe Use of Chemicals

The problem of information of documents is compounded by the fact that there is so much information in system and software specifications, while only a small subset of it may be relevant in any given context (Leveson 2000). Risk management should be considered as a control function focused on maintaining a particular hazardous, productive process within its boundaries of safe operation and that a systems approach based on control theoretic concepts should be applied to describe the overall system functions (Rasmussen 1997). Systems are viewed as interrelated components that are kept in a state of dynamic equilibrium by feedback loops of information and control (Dulac and Leveson 2004). The management of risk requires that there is a comprehensive, structured plan in place with clear responsibilities and authority, which consistently reduces risk to an acceptable level that can be tolerated by all concerned and insures that the integrity of risk reduction is maintained throughout the life of the process (Mostia 2009).

In a learning organization, leaders are designers, stewards and teachers (Senge 1994, p. 340). They are responsible for building organizations where people continually expand their capabilities to understand complexity, clarify vision and improve shared mental models—that is, they are responsible for learning. As design decisions are made, the analysis is used to evaluate those decisions and their effects on the system safety (Dulac and Leveson 2004). The system design must ensure that the safety constraints continue to be enforced as changes occur throughout the life of the system. The control processes that enforce these constraints must limit system behaviour to the safe changes and adaptations implied by the constraints (Dulac and Leveson 2004). van Asselt and Renn (2011) stated that "risk governance" pertains to the various ways in which many actors, individuals, and institutions, public and private, deal with risks surrounded by uncertainty, complexity, and/or ambiguity". Furthermore, it refers to the totality of actors, rules, conventions, processes, and mechanisms concerned with how relevant risk information is collected, analysed and communicated, and how regulatory decisions are made.

2.3 Measures of Management

The results of Le Coze (2013) indicated that an organization maintains OSH through implementation of all the leadership activities (e.g. training and learning from experience). Grote (2012) stated that the move towards goal-oriented regulation can also be understood in the context of the general trend towards acknowledging the need to cope with uncertainty instead of trying to manage it out of existence. In view of the increasing complexity of socio-technical systems the concept of anticipation and perfect prediction as basis for legal regulation becomes less feasible. Hale and Swuste (1998) introduced a useful distinction between different types of rules. They distinguished three types, which differ in the degree of freedom they provide the individual adhering to regulations different way to undertake their exact behaviour: (1) Performance goals, which define only what has to be achieved and not how it must be done; (2) Process rules, which define the process by which the person or organization should arrive at the way they will operate, but still leave considerable freedom about what that operation will be; and (3) Action rules, which specify in terms of 'If—Then' statements exactly how people shall behave, or how hardware (e.g. personal protective devices) shall be used at that moment in time.

An understanding of the scenarios and risks in the process hazard analyses can be used to establish highly effective aids to decision making by the management (Myers 2013). Policies that demonstrate management's commitment to OSH emphasize to the workers that they are supported by the organization (Dejoy et al. 2010). Olsen (2010) found that top-level management is primarily responsible for defining the correct OSH goals and signalling the importance in contributing to other organizational goals.

When the product or process is put to use, best practices for implementing the safe use procedures should be followed and any residual risks should be promptly identified and evaluated (Baker et al. 2007; Baram 2007). Finally, the residual risks need to be addressed by remedial measures, such as design change and improved safe use procedures. van Asselt and Renn (2011) created a definition of "risk governance" as a horizontally organized structure of functional self-regulation encompassing state and non-state actors bringing about collectively binding decisions without superior authority hints at treating governance in terms of a normative model (e.g. self-regulation) for how societal decision-making should be organized. A typical industrial process is a complex and dynamic system in which the safety level of the system (or industrial plant) will vary with time and therefore should be monitored continually and the system reliability constantly updated (Pasman et al. 2009). This reliability approach should start by incorporating elements of inherent safety and resilience at the design stage (e.g. Hollnagel et al. 2006).

2.4 Hypotheses 1–3

Hypothesis 1. The aggregated variable "Prevention prioritizations" is positively related to the aggregated variable "Safe use of chemicals".
Hypothesis 2. The aggregated variable "Measures of management" is positively related to the aggregated variable "Safe use of chemicals".
Hypothesis 3. The aggregated variable "Prevention prioritizations" is positively related to the aggregated variable "Measures of management".

3 Data Collection, Methods and Participants

3.1 The Quantitative Study: Online Questionnaire

The online questionnaire research was conducted in October, 2008. The study method consisted of a questionnaire form distributed via e-mail and available over the internet. The questionnaire was filled in by both employers and workers. The companies of the employers' respondents were selected from the Responsible Care Company register of the Chemical Industry Federation of Finland. The worker respondents concerned OSH representatives of different companies were selected from the registers of the Finnish Chemical Workers Trade Union (workers) and of the Union of Salaried Workers (clerical workers). The questionnaire was sent to 108 employers' representatives (HSE managers), of whom 49 responded, resulting in a response rate of 45 %. The respondents to the employers' questionnaire were from 45 different companies. The questionnaire was sent to 217 workers' OSH representatives of, of whom 105 responded, resulting in a response rate of 48 %. The respondents to the workers' questionnaire were from 80 different companies. The questionnaire form is being presented as the appendix of this article. Respondents were asked to indicate how the different topics were executed, ranging from 1 (very good), 2 (rather good), 3 (poorly), 4 (not executed). Statistical analysis was done with SAS software (2005). In testing the data regression analysis was applied.

3.2 The Qualitative Study: Discourses of the Employers and Workers

In the data analysis of the discussion in the on-line questionnaire, the actantial model of Greimas (1987, 1990) was applied.

The six actants are divided into three oppositions, each of which forms an axis of the actantial description (Hébert 2011):

- The axis of goal: (1) subject/(2) object. The subject is what is directed towards an object. The relationship established between the subject and the object is called a junction.
- The axis of power: (3) helper/(4) opponent. The helper assists in achieving the desired junction between the subject and object; the opponent hinders this goal.
- The axis of transmission and knowledge: (5) sender/(6) receiver. The sender is the element requesting the establishment of the junction between subject and object. Sender elements are often receiver elements as well.

The core of the actantial model is formed from the subject's intention to acquire the object. Consequently, the subject is the performer of a task, and the object is something that the subject tries to acquire through some activity (Greimas and Courtes 1982). A sender is a party who gives the task to the subject (Greimas 1987, 1990).

At the same time, the sender defines the value targets for the activity, and provides justification and motivation for the subject's activity. A receiver is a party who evaluates whether the subject has succeeded, and confers on the parties to the discourses either a reward or punishment after the task has been completed.

In addition, the actantial model (Greimas 1987, 1990) includes so-called secondary actors: a helper and opponent. The helper promotes the subject's ability to perform the task. The importance of the role of the opponent is emphasized when it represents a counterforce that the subject has to overcome in order to reach its goal. When carrying out actantial reading, the researcher is trying to identify the following matters in the text being analysed: Who is the main character (subject) of the discourse (or of the issue to be dealt with)? What does the subject try to acquire (the object)? Who defines the subject's obligation to acquire the object, or justifies and motivates the subject to acquire it (sender)? Who helps the subject when he/she tries to carry out the task (helper)? Who opposes the subject when he/she tries to acquire the object (opponent)? Who obtains an advantage or disadvantage when the subject acts (receiver)?

4 Questionnaire Survey Results

4.1 Testing of the Hypotheses

4.1.1 Regression Analysis Among the Aggregated Variables

For workers, in the regression analysis both aggregated variable "Prevention prioritizations" and aggregated variable "Measures of management" had a highly significant ($p < 0.001$) effect on aggregated variable "Safe use of chemicals" (Table 1).

Similarly for employers, in the regression analysis both aggregated variable "Prevention prioritizations" and aggregated variable "Measures of management" displayed a highly significant ($p < 0.001$) effect on aggregated variable "Safe use of chemicals".

In addition, for workers, in the regression analysis both aggregated variable "Prevention prioritizations" and aggregated variable "Safe use of chemicals" had a highly significant ($p < 0.001$) effect on aggregated variable "Measures of management" (Table 2).

Similarly for employers, in the regression analysis both aggregated variable "Prevention prioritizations" and aggregated variable "Safe use of chemicals" showed a highly significant ($p < 0.001$) effect on aggregated variable "Prevention measures of management".

Table 1 For workers, regression analysis of the importance of positive impacts of legislation on aggregated variable "Safe use of chemicals" and a set of aggregated variable "Prevention prioritizations" and "Measures of management": dependent aggregated variable: "Safe use of chemicals"

Source	df	Sum of squares	Mean square	F value	Prob. level	R-square
Model	1	2,745	2,745	97.30	<0.0001	0.49
Error	102	2,878	28.22			
Corrected total	103	5,624				

Parameter estimates

Independent variable	df	Parameter estimate	Standard error	t-value	Prob. level	95 % confidence limit	
Intercept	1	12.64	2.64	4.80	0.0001***	7.42	17.87
Prevention prioritizations	1	0.69	0.07	9.86	0.0001***	0.40	0.757
Intercept	1	24.54	2.13	11.53	0.0001***	20.29	28.74
Measures of management	1	0.47	0.07	6.69	0.0001***	0.33	0.60

$p < 0.001$***

Table 2 Regression analysis for the workers assessments for importance of positive impacts of legislation on aggregated variable "Measures of management" and a set of aggregated variable variables "Prevention prioritizations" and aggregated variable "Safe use of chemicals": dependent aggregated variable: "Measures of management"

Source	df	Sum of squares	Mean square	F value	Prob. level	R-square
Model	1	2,404	2,404	44.75	<0.0001	0.31
Error	102	5,479	53.71			
Corrected total	103	7,883				

Parameter estimates

Independent variable	df	Parameter estimate	Standard error	t-value	Prob. level	95 % confidence limit	
Intercept	1	2.08	3.29	0.63	0.53	−4.45	8.61
Prevention prioritizations	1	0.74	0.09	8.38	0.0001***	0.57	0.92
Intercept	1	4.28	3.80	1.13	0.2	−3.25	11.81
Measures of management	1	0.65	0.10	6.69	0.0001***	0.46	0.85

$p < 0.001$***

4.1.2 Results of the Testing of H1–H3

The hypotheses (H1–H3) were supported.

4.2 The Collective OSH Climates

4.2.1 The Climate Factors for "Safe Use of Chemicals"

For workers, the factor analysis on "Safe use of chemicals" was being carried out. In the rotated factor analysis according to the variables three factors were detected (Table 3).

Standardized Cronbach Coefficient Alpha is 0.89. Kaiser–Meyer–Olkin measure of sampling adequacy is 0.82. According to the variables, Factor 1 is named as

Table 3 Workers' climate for rules compliance in "Safe use of chemicals"

Safe use of chemicals	Factor 1	Factor 2	Factor 3	Communality	Weight
(1) On-the-job training of subcontractors working at the workplace is arranged	**0.75**	0.27	0.28	0.71	3.53
(2) Sub contractors carry out danger identification in the same way as the client enterprise	**0.71**	0.18	0.26	0.61	2.54
(3) On-the-job training is arranged	**0.56**	0.33	0.23	0.48	1.94
(4) Adequacy of the preventive measures is assessed	**0.55**	0.33	0.34	0.53	2.13
(5) Accident dangers are assessed	**0.54**	0.37	0.37	0.56	2.28
(6) Exposure to chemical substances are carried out	**0.49**	0.40	0.33	0.51	2.05
(7) Work-hygienic measurements are carried out	**0.48**	0.44	0.02	0.43	1.76
(8) Policy of OHS has been formulated	0.22	**0.81**	0.17	0.73	3.76
(9) Follow up and reporting is regular	0.38	**0.80**	0.16	0.80	5.09
(10) OHC service makes work place assessments	0.28	**0.58**	0.38	0.56	2.26
(11) Chemicals are stored safely	0.21	0.22	**0.81**	0.74	3.86
(12) Chemicals are used safely	0.28	0.10	**0.78**	0.70	3.33
(13) Chemicals safety data sheets are utilised	0.28	0.10	**0.78**	0.43	1.75

"Training", Factor 2 "OSH policies" and Factor 3 "Work processes". For workers, variance explained by each factor is as follows: Factor 1–30 %, Factor 2–27 %, and Factor 3–22 %.

4.2.2 The Climate Factors for "Prevention Prioritizations"

Factor analysis of the workers on "Prevention prioritizations" is presented in the Table 4.

Table 4 Workers' climate for rules compliance in "Prevention prioritizations"

Prevention prioritizations	Factor 1	Factor 2	Factor 3	Communality	Weight
(1) Dangers are eliminated	**0.78**	0.18	0.18	0.68	3.10
(2) Collective safety measures which have a general impact are adopted before individual measures	**0.74**	0.24	0.10	0.61	2.57
(3) Worker expertise and experience are utilized in connection with danger identification and risk assessments	**0.64**	0.35	0.28	0.61	2.58
(4) Development of technology is taken account	**0.62**	0.35	**0.42**	0.69	3.21
(5) Dangerous substances are replaced by less dangerous	**0.51**	0.26	0.19	3.63	1.57
(6) In danger identification and risk assessments OHC providers are used as experts in risk assessments	**0.47**	0.38	0.25	0.42	1.72
(7) Risks are assessed	0.24	**0.76**	0.18	0.67	3.06
(8) Documents of the risk assessment are up-dated	0.24	**0.74**	0.14	0.63	2.70
(9) Dangers are identified	0.25	**0.67**	0.30	0.60	2.49
(10) Documents of the risk assessment are kept available to workers	0.20	**0.64**	0.12	0.46	1.84
(11) Danger identification and risk assessments are carried out in cooperation between the employer and workers	**0.42**	**0.59**	0.21	0.56	2.29
(12) Continuous monitoring of the working environment is taken care	0.25	0.30	**0.92**	0.16	**0.92**
(13) Effects of preventive measures are followed-up	**0.50**	0.23	**0.58**	0.63	2.73

Standardized Cronbach Coefficient Alpha is 0.91. Kaiser–Meyer–Olkin measure of sampling adequacy is 0.84. According to the variables, Factor 1 is named as "Technology", Factor 2 "Assessments and documents" and Factor 3 "Control". For workers, the variance explained by each factor are as follows: Factor 1: 31 %, Factor 2: 30 % and Factor 3: 18 %.

4.2.3 The Climate Factors for "Measures of Management"

Factor analysis of the workers on "Measures of management" is presented in the Table 5.

Standardized Cronbach Coefficient Alpha is 0.88. Tucker and Lewis's Reliability Coefficient is 0.93. Kaiser–Meyer–Olkin measure of sampling adequacy is 0.91. According to the variables, Factor 1 is named as "Leadership", Factor 2 "Assessments" and Factor 3 "Utilization of measures". For workers, variance explained by each factor are as follows: Factor 1: 24 %, Factor 2: 23 % and Factor 3: 23 %.

4.3 Qualitative Results of the Questionnaire Survey: Discourse Analysis

A semi-structured format was used in the questionnaire concerning open responses. Respondents were asked to comment on the following topics: (1) the development need and challenges of the legislation, (2) the need of information concerning EU regulation REACH and (3) the cooperation with different Governmental Inspectorates in chemical safety.

4.3.1 The Development Need and Challenges of the Legislation

Legislation was considered as functionally good but clearer application instructions were desired.

- "The actual legislation is extensive but employers' obligations to comply with the legislation should be emphasised." (Worker representative)
- "Risk assessments should be handled by the same act regardless of whether they relate to risks relating to OSH risk or to risks associated with the environment of the factory." (Employer representative)
- "The law should define the frequency of risk assessments." (Worker representative)
- "It is often thought that risk assessment issues are in order when the risks of the production are assessed." (Employer representative)

Table 5 Workers' climate for rules compliance in "Measures of management"

Measures of management	Factor 1	Factor 2	Factor 3	Communality	Weight
(1) Results of the risk assessments are utilized when planning workstations	**0.79**	0.18	**0.43**	0.84	6.34
(2) Results of the risk assessments are utilized when following up the improvements of the working environment	**0.73**	**0.44**	0.25	0.79	4.82
(3) Results of the risk assessments are utilized in planning and leadership of the work	**0.59**	**0.48**	0.33	0.68	3.13
(4) Results of the risk assessments are utilized when developing the work of the supervisor	**0.55**	**0.40**	0.26	0.54	2.15
(5) Results of the risk assessments are utilized when planning the clarifications and measurements	0.36	**0.77**	0.35	0.84	6.14
(6) Results of the risk assessments are utilized when drawing up instructions and guidelines for use	0.32	**0.70**	**0.44**	0.79	4.75
(7) Results of the risk assessments are utilized in the on-the-job training of the workers	0.34	**0.62**	**0.46**	0.72	3.63
(8) Results of the risk assessments are utilized in the realization of the policy of OHS	**0.46**	**0.51**	**0.42**	0.65	2.88
(9) Results of the risk assessments are utilized when renovating industrial premises	0.29	0.30	**0.72**	0.69	3.23
(10) Results of the risk assessments are utilized when planning and executing the activities of the OHC	0.26	0.34	**0.68**	0.65	2.82
(11) Results of the risk assessments are utilized in the acquisition of machines and devices	0.36	0.34	**0.59**	0.59	2.41
(12) Results of the risk assessments are utilized when planning and implementing development preventive measures	**0.49**	0.39	**0.54**	0.68	3.17

- "Risk assessments should be included in project activities, internal service-providing activities and maintenance. OSH legislation should somehow emphasize these areas which would make it easier to justify the importance here in the field. Of course these risk assessments can be based on something else than OSH legislation. It is a fact that risks are even bigger during maintenance and servicing work, which are often carried out while the production is still going on." (Employer representative)
- "Risk assessment in SMEs is difficult. There is a need to provide guidelines to SMEs for carrying out the risk assessments. We use the HSEQ assessment system. We use our own resources for guiding SMEs (small and medium-size enterprises)." (Worker representative)
- "It should be guiding and produce comparable risk assessment both for authorities and enterprises." (Employer representative)
- "There should be more assessment of joint effects." (Worker representative)
- "Training and enough time should be arranged to allow oneself to become familiar with OSH issues. It would be necessary to go through appropriate exemplary cases so that all persons involved would understand how important this is. In many workplaces some issues like noise are not taken seriously." (Worker representative)
- "I think the legislative basis is quite good. However, more practical guidance is needed in its interpretation." (Employer representative)
- "The legislation is ok as far as it is interpreted openly e.g. in connection with authority inspections, taking account of the field of industry and other circumstances. With regard to the development of operation and safety, the same model is not suitable or at least not useful for all enterprises." (Employer representative)

The above-mentioned comments in the Greimas' (1987, 1990), Greimas and Courtes (1982) semiotic actantial model linking the theory with the textual comments which are referring to the "Legal demands with respect to risk assessment and prevention measures" are presented in Fig. 2.

4.3.2 The Need of Information Concerning EU Regulation REACH

The respondents were also asked to answer open questions about the challenges the new EU chemical regulation REACH sets for enterprises in the future. Some of the responses emphasized that issues relating to REACH shall be taken care of in a centralized manner, and part of the responses were clearly doubtful. Uncertainty about the REACH regulation was very common, especially in the responses of worker representatives.

- "REACH will mainly be taken care of by others—those at the supplier and management level of the concern. The greatest challenge could be that we need to collect information from the downstream users of our products." (Employer representative)

11 Safe Use of Chemicals and Risk Prevention …   173

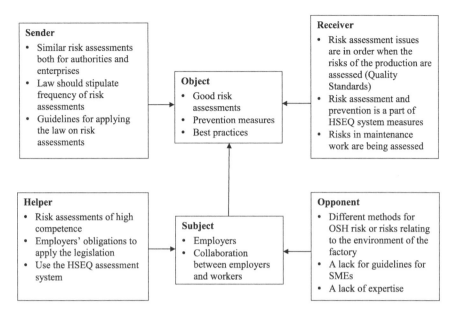

Fig. 2 The actantial model as it applies to the "legal demands with respect to risk assessment and prevention measures"

- "The competence level of documents and descriptions regarding exposure and use should be raised. At some point the deliveries of small lots of chemicals may stop because of excessive registration costs, which may cause problems for taking into use new substituting chemicals." (Employer representative)
- "REACH will be taken care of at the division level. I have not received much information about it from that direction. However, I have found out for myself that at least the obligations of final users also apply to us." (Employer representative)
- "The work will be carried out in a centralized manner by resources of the concern." (Employer representative)
- "The whole REACH concept is very unclear." (Employer representative)
- "I cannot comment on issue that I don't know well enough. Does this mean that more information should be given on this?" (Worker representative)
- "I am worried that the knowledge and skill of the authorities might not be good enough to enforce the provisions of the REACH regulation at enterprises on local level." (Employer representative)
- "The principles of risk assessments should also be followed when developing requirements. REACH seems to decentralise the resources, which leaves too little resources for controlling real risks. Issues representing low risk levels can demand an unreasonable amount of the resources. Surveys and assessments are made in a very burdening way." (Employer representative)
- "The regulation has remained quite distant from the level of everyman." (Worker representative)

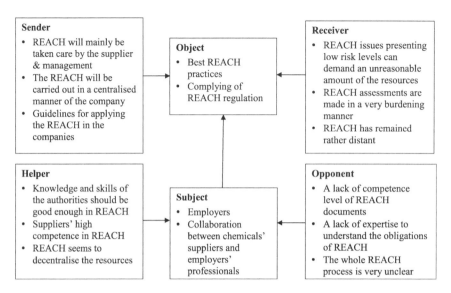

Fig. 3 The actantial model as it applies to the "need of information concerning EU regulation REACH"

- "I don't know the regulation well enough to be able to respond." (Worker representative)
- "First it should be explained to the workers what the REACH regulation of the EU actually means. Only a few people know something about that regulation so it is difficult to have an opinion." (Worker representative)

The above-mentioned comments in the Greimas' (1987, 1990), Greimas and Courtes (1982) semiotic actantial model linking the theory with the textual comments which are referring to the "Need of information concerning EU regulation REACH" are presented in Fig. 3.

4.3.3 The Cooperation with Different Governmental Inspectorates in Chemical Safety

The respondents could comment via open questions were asked to describe how cooperation between different authorities should be developed so the enterprises could better integrate the risk assessment obligations rising from the legislation. For example, the respondents wrote that the activities of the different authorities should be undertaken in a uniformed manner.

- "The various authorities (e.g. Finnish Environment Institute, the regional OHS governmental authorities) could consider a joint risk assessment procedure. This would mean that every enterprise could assess all risks simultaneously, taking account of all sectors of its activity. In addition, joint inspections could be

arranged instead of having several inspections per year, arranged by various authorities, as the situation is now." (Employer representative)
- "This sector should be developed so all authorities could be served by the same risk analysis." (Employer representative)
- "The regional OSH authorities and civil protection authorities should cooperate. Some issues overlap in the inspections of these authorities. A fire marshal and an occupational safety inspector are both competent to notice untidy environments endangering safety. However their opinions may be different." (Worker representative)
- "Combining the Acts could bring about a similar interest in all risks involved. Now separate mappings and examinations are carried out which do not necessarily take into account the effects of the risks as a whole." (Worker representative)
- "An integrated risk analysis would provide opportunities to focus on the most important issues in the particular enterprise in question, regardless of whether the issues relate to the environment or occupational safety and/or major accident risks. An integrated risk analysis would also create a good basis for defining the right grade of involvement of the different authorities." (Employer representative)
- "I don't think that one integrated risk analysis can suit perfectly for assessing all risks. It would be great if they succeeded in finding a model that would serve several purposes at the same time." (Employer representative)
- "At our workplace this is a continuum that is carried out under the action plan. Audits guarantee that progress happens all the time. I hope that the authorities do not become more involved because people are very well aware of these issues." (Worker representative)
- "The number of inspections carried out by regional OSH authorities should be increased." (Worker representative)
- "More resources are needed for the enforcement of OSH issues and for workplace visits of inspectors. The inspections are useful for all parties." (Employer representative)
- "Inspectors of the regional OSH authorities should check the real situation at workplaces. Instructions should be given to the employer, and corrective measures should be followed-up." (Worker representative)
- "Referring to the previous issue also standardized management systems are fixed to different sectors. An integrated model would integrate all safety risks at enterprises basically to the same level. Under authority control, any voluntary certifications of management systems could be taken account of better than now." (Employer representative)

The above-mentioned comments in the Greimas' (1987, 1990), Greimas and Courtes (1982). Greimas semiotic actantial model linking the theory with the textual comments which are referring to the "Cooperation with different Governmental Inspectorates in chemical safety" are presented in Fig. 4.

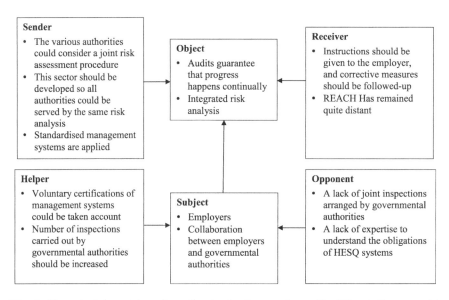

Fig. 4 The actantial model as it applies to the "cooperation with different Governmental Inspectorates in chemical safety"

5 Discussion

5.1 Theoretical and Practical Implications

For "Safe use of chemicals", Factor 1 is named as "Training", Factor 2 "OSH policies" and Factor 3 "Work processes". This present study supports the findings of Grote (2012) that the predominant approach to managing uncertainty in any particular organization needs to be taken into account and incorporated into the design, implementation as well as when evaluating safety management systems. The present study when asked for comments about "Training" supports the conclusions of Brennan and Kiihne (2009) that a good training and process safety information system is one that is reviewed periodically. In terms of personal safety, training is mainly aimed at increasing each individual's knowledge about risks and risk protection measures, which today can be found in almost all industrial sectors (Grote 2012). This study confirms the results of Baram (2007) that companies must be alert to existing and pending regulations that can have direct and indirect implications about the design of their products and processes. The present results in the section of "OSH policies" support the conclusions of Helland (2009) that in high-risk industries, complex and novel tasks require more goal and process policies in order to promote exploration and learning. Many companies in high-risk industries have started to invest in team training in order to develop the skills needed for effective coordination (Grote 2012). Feedback is more influential than simple training and is enhanced if there is direct dialogue between the employer and the workers (Hale et al. 2010). In the section evaluating "Work processes" this study indicated that the employer shall ensure that the risk is reduced to a minimum by application of protection and prevention measures e.g. with a design of appropriate work processes, engineering controls and use of adequate equipment and materials. In addition, the employer shall establish procedures (action plans) which can be put into effect when any an accident or incident such event occurs. The present results support findings of Grote and Künzler (2000) who stated that "Efforts aimed at increasing the safety of high-risk production systems focus more and more not only on technical and individual-centered measures, but on an integral safety management improving the interplay of technology, organization and human resources". Hale et al. (1998) established and tested the management of safety in maintenance activities in the chemical process industry and developed an audit checklist to carry out in-depth assessment of their management systems. Reniers (2009) noted that due to the rapid systems development of chemical technology, in chemical plants there has been is a continuous growth of ever more complex installations operating under more extreme and critical process conditions. Schöbel and Manzey (2011) proposed that OSH efforts into high hazard systems should focus more strongly on the development of alternative models of socio-technical systems' performance; the learning from mistakes has the potential to supplement existing methods of functional event analysis.

For "Prevention prioritizations", Factor 1 is named as "Technology", Factor 2 "Assessments and documents" and Factor 3 "Control". In this section assessing risk assessment, this study supports in the findings of Grote (2012) that in order to draw up useful procedures in risk prevention, it is important to look beyond the general level of procedures into different types of rules and how they affect organizational processes. The present results in the section of "Technology" support the conclusions of Helland (2009) that the good design of a manufacturing process reduces or eliminates the hazards associated with materials and operations. Furthermore, this can be done in several ways (a) minimize-use smaller quantities of hazardous substances; (b) substitute-replace a material with a less hazardous substance; (c) moderate-use less hazardous conditions, a less hazardous form of a material, or facilities which minimize the impact of a release of hazardous material or energy; (d) simplify-design facilities which eliminate unnecessary complexity and make operating errors less likely (Helland 2009). Wu et al. (2011) emphasized the system importance of the safety leadership highlighting that safety climate mediated the relationship between safety leadership and performance. A participation by employees in the leadership of their organizations may imply that they accept, or show readiness to accept, work roles which go beyond the agreements and constraints evolved by negotiation between management and workers on their behalf (Cherns 1976).

This present study supports in the section of "Assessments and documents" the results of Hale et al. (1998) that information on the results of risk assessment, especially in maintenance work, should receive attention with respect to the safety measures in high-risk industry. This study supports the criteria presented by Neathey et al. (2006, p. 44) i.e. that that there are good ways to inform workers of the results of their risk assessments and for communicating OSH issues as an agenda item at all staff meetings. The present results support the findings of Payne et al. (2009) that it is very important that the workers participate in the risk assessment in all aspects of process safety. If the organization is not communicating the risks of the job to its workers nor using previous unsafe incidents as learning opportunities, then workers will not be able to develop the skills to act safely, recognize and report dangers, and correct problems as they arise (Payne et al. 2009). This study is for support the conclusion of Gadd et al. (2004); here in the section of "Control" this study emphasized the importance of informing about decisions relating to the control and reduction of workplace risks. This study confirms the results of Gadd et al. (2004) that risk assessment process can be divided into three stages—preparing for the assessment, carrying out the assessment and post-assessment activities. The results found in the section in "follow-up and monitoring" the support to findings of Hale and Swuste (1998) they established three levels of safety management systems as the goal monitoring, follows-up procedures, and execution of work instructions.

For "Measures of management", Factor 1 is named as "Leadership", Factor 2 "Assessments" and Factor 3 "Utilization off measures". In the section assessing prevention measures, this study supports the findings of Hale and Swuste (1998) who stated that management may be advised to set goal, process, and action rules. First, the goal rules define the OSH goal to be achieved, leaving open how this is accomplished by the actors concerned. Second, the process rules provide guidance for deciding on the right course of action for achieving certain goals. Third, the action rules prescribe detailed courses of action, possibly without even mentioning the goal to be achieved. Action rules should be used when stability of processes is required. Goal and process rules should be used when flexibility is required. The employer shall take the necessary preventive measures e.g. hazardous chemical agents should be eliminated or reduced to a minimum by the design and organization of systems of work at the workplace. This study supports in the section of "Leadership" the results of Baram (2007) that regulations also have important indirect effects on leadership and process design. Furthermore, Hale et al. (2007) concluded that the organization of OSH involves the use of a combination of goal-oriented risk analyses with the participation of experienced users to arrive at detailed design solutions. The present results found in the section of "Assessments" support to the conclusions of Mohaghegh and Mosleh (2009) that the four activities are being included as "design", "implementation", "internal auditing" and "change", which are similar to the cycle 'plan, do, check and act', which has its roots in the quality control field. The present results in the section of on "Utilization off measures" support the findings of Helland (2009), who postulated that pursuing the concept of inherency and its related risk may provide an effective orientation of the potential of a substance, product, process or system to evoke OSH harm.

The present qualitative results support those of Grote (2012) that managing uncertainty is an important feature also in general organizational characteristics. In addition, Grote (2012) classified the focus of OSH management as follows: (1) Tasks and work processes—e.g., task complexity, types of task interdependencies;

(2) People: i.e. do they have formal qualifications; (3) Organization structure, i.e. the distribution of decision authority; (4) Technology: i.e. technical work processes; (5) External relationships: i.e. supply chain, cooperation between the employers and governmental authorities. The qualitative results support with respect to EU REACH Regulation the conclusions of van Asselt and Renn (2011) that the decisions about chemicals risks are taken in complex webs of actors, rules, conventions, processes, mechanisms, institutional arrangements, and political cultures. Risk decisions of REACH regulations can be understood only as the outcome of complex interplays between multiple actors. In topics related to REACH this study can be considered to be in agreement with the conclusions of van Asselt and Renn (2011) that a governance perspective is needed to examine and explain the societal dynamics around issues framed as risk issues.

With respect to REACH this study confirms the results of van Asselt and Renn (2011) that in supply chains the systemic risks requires a holistic hazard identification, risk assessment, and risk management because investigating systemic risks goes beyond the usual agent-consequence analysis.

This study showed that for workers and for employers, in the regression analysis both "Risk assessments process and prevention prioritizations" and "Prevention measures of management" showed a highly significant ($p < 0.001$) effect on "Safe use of chemicals". With respect to risk assessment process this study supports the results of van Asselt and van Bree (2011) that from the perspective of 'risk governance', the precautionary principle should be considered as the obligation to take uncertainty seriously in all stages and in all dimensions and furthermore becoming a normative, instead only of a legal principle. The present results support the findings of Carmeli et al. (2013), who indicated that if one has a resilient top management then good communication within top management teams facilitates engagement in strategic decision comprehensiveness. Furthermore, Carmeli et al. (2013) found that managers should employ more comprehensive and extensive decision making processes e.g. by improving their level of strategic grasp of their environments and processing information for knowing how to manage complex processes. This study supports the results of Rasmussen (1997) that it is evident that in high-risk industries safety control requirements and a new approach to representation of system behaviour should clearly be focused on the mechanisms generating behaviour in the dynamic work context. In addition, risk assessment should become an integral part of a company's safety management system. This kind of assessment will almost inevitably result in recommendations for improvements and further actions to control and reduce risk. The present results support previous reports of Gadd et al. (2004) that risk assessment should not be a one off activity, but should be ongoing and be a part of the process of continuous improvement. Thus, the risk assessments need to be reviewed as part of the standard OHS management practice. The present results confirm the conclusions of Hale and Swuste (1998) that with respect to on-the-job training a complex task requires more goal and process rules in order to promote learning. This study supports the findings of Grote (2012) that different types of OSH regulation regimes impose very different demands on their internal functioning (Grote 2012). When the focus is on external regulation, companies have to continuously react to changes of regulatory demands and adapt their safety management accordingly. All companies, especially those operating with major risks are subject to some degree of external regulation. For companies operating in high hazard systems, the different types of OSH regulation regimes demand levels of excellence in their internal functioning.

Recent years have witnessed a tendency to move away from prescriptive regulation which specifies in minute detail how an organization should carry out its operations, and nowadays the trend is to goal-oriented legislation, which promotes self-regulation (Grote 2012). Furthermore, in particular during a period with a rapid pace of change, it is important to analyze how effectively information of changes of technology, processes, and policies are being communicated (Rasmussen 1997).

In this approach the system thinking would be of great importance. The systems thinking does not mean simply ignoring the complexity inherent in industrial processes, but instead it involves organizing complexity into coherent measures that illuminate the causes of problems and how they can be remedied in an enduring manner (Senge 1994, p. 128). Human factors (ergonomics) approach is a systems discipline and profession, applying a systems philosophy and systems approaches (Wilson 2014). These concern (Edwards and Jensen 2014): (1) Boundaries and scope of the system, (2) Participants in a system, (3) The character of knowledge, (4) Performance management, leadership and key performance indicators. Väyrynen et al. (2014) applied a holistic ergonomics/human factor approach to optimize work systems, as far as performance and effectiveness, including in a key role people without detriment to their health, safety, or other factors of well-being at work. Optimization may be evaluated based on measures of three categories: (a) health and well-being, (b) safety and (c) production performance (the quantity and quality (Q) of production) with minimal nonconformities (Väyrynen et al. 2014).

5.2 Limitations

Some limitations should be considered when interpreting the present results. This kind of questionnaire does not allow one to draw causal conclusions with complete confidence. The present results are subject to potential inaccuracies related to the inability of respondents to recall information correctly. First, the main limitation of the current study is its cross-sectional nature. Second, our study relies exclusively on self-report measures. Third, the cross-sectional nature of the survey did not allow for the determination of causal relationships. Fourth, the generalizability of our findings should be interpreted with caution, as the study was conducted in the realm of the Finnish chemical industry. Fifth, longitudinal studies are needed to establish the causality of the relationships. These limitations notwithstanding, this study provides a new kind of the perspective on practically orientated recommendations that are linked to the distinctive contexts of work places of Western Europe. Longitudinal designs and experimental intervention are needed to make it possible to draw more convincing causal inferences in OSH.

Limitations concerning the sample of the participants need to be considered and this limits the generalizability of the results. At the same time, the statistical sample size of the small and micro enterprises (1–9 workers) is under-represented.

6 Conclusions and Recommendations

In the regression analysis of workers and for employers, two aggregated variables i.e. "Prevention prioritizations" and "Measures of management" exerted a highly significant effect on one aggregated variable "Safe use of chemicals". For the workers, in the climate analysis for "Safe use of chemicals" the significant factors were as follows: Factor 1—Training, Factor 2—OSH policies and Factor 3—Work processes. In the climate analysis for "Prevention prioritizations" the significant factors were as follows: Factor 1—Technology, Factor 2—Assessments and documents and Factor 3—Control. In the climate analysis for "Measures of management" the following factors were important: Factor 1—Leadership, Factor 2—Assessments and Factor 3—Utilization of measures.

The risk assessment should not be a one-off activity, but should be ongoing and a part of the process of continuous improvement. Information on the results of risk assessment should be communicated to all interested partners, for example as an item of the agenda at all staff meetings. In OSH management, certain measures should be applied proactively e.g.: (1) in the implementation of risk assessment in the planning and leading of the work, (2) in the development of the supervisors' leadership and (3) in the following up of the improvements of the working conditions. The management should make every effort to ensure that workers have the opportunity to participate in prevention of safety-related pitfalls, e.g. problems discovered in the risk assessment. In risk assessment, the following measures should to be implemented e.g.: (1) documents concerned with risk assessment should are be up-dated, (2) relevant documents should be made available to the workers, (3) hazardous substances should be replaced by less hazardous alternatives and (4) effects of preventive measures are being followed-up.

Acknowledgments The author is grateful for help with the statistical analysis of the results to Maria L. Hirvonen from the Finnish Institute of Occupational Health.

References

Baker III JA, Bowman FL, Erwin G, Gorton S, Hendershot D, Leveson N, Priest S, Rosenthal I, Tebo PV, Wiegmann DA, Wilson LD (2007) The report of the BP U.S, refineries independent safety review panel. The baker report. http://www.bp.com/liveassets/bp_internet/globalbp/STAGING/global_assets/downloads/Baker_panel_report.pdf. Accessed 16 June 2014

Baram M (2007) Liability and its influence on designing for product and process safety. Saf Sci 45 (1–2):11–30

Brennan J, Kiihne GM (2009) A little knowledge is a dangerous thing—unexpected reaction case studies make the case for technical discipline. J Loss Prev Process Ind 22(6):757–763

BS 8800 (2004) Occupational health and safety management systems-guide. British Standard Institute, London

Carmeli A, Friedman Y, Tishler A (2013) Cultivating a resilient top management team: the importance of relational connections and strategic decision comprehensiveness. Saf Sci 51 (1):148–159

Cherns A (1976) Principles of socio-technical design. Hum Relat 29(8):783–792. http://hum.sagepub.com/content/29/8/783.full.pdf+html. Accessed 18 June 2014

DeJoy DM, Della LJ, Vandenberg RJ, Wilson MG (2010) Making work safer: testing a model of social exchange and safety management. J Saf Res 41(2):163–171

Dulac N, Leveson N (2004) An approach to design for safety in complex systems. In: International conference on system engineering (INCOSE '04), Toulouse, June 2004, 14 pp. http://sunnyday.mit.edu/papers/incose-04.pdf. Accessed 16 June 2014

Edwards K, Jensen PL (2014) Design of systems for productivity and wellbeing. Special issue: systems ergonomics/human factors. Appl Ergon 45(1):26–32

EU Directive (1998) EU Directive 98/24/EC of 7 April 1998 on the protection of the health and safety of workers from the risks related to chemical agents at work (fourteenth individual Directive within the meaning of Article 16(1) of Directive 89/391/EEC). Official Journal L 13, 5.5.1998, pp 11–23. http://eurlex.europa.eu/LexUriServ/LexUriServ.do?uri=CELEX:31998L0024:EN:HTML. Accessed 18 June 2014

Gadd SA, Keeley DM, Balmforth HF (2004) Pitfalls in risk assessment: examples from the UK. Saf Sci 42(9):841–857

Griffin MA, Neal A (2000) Perceptions of safety at work: a framework for linking safety climate to safety performance, knowledge, and motivation. J Occup Health Psychol 5(3):347–358

Greimas AJ (1987) On meaning: selected writings in semiotic theory. University of Minnesota Press, Minneapolis

Greimas AJ (1990) The social sciences: a semiotic view. University of Minnesota Press, Minneapolis

Greimas AJ, Courtes J (1982) Semiotics and language: an analytical dictionary. Indiana UP, Bloomington

Grote G (2012) Safety management in different high-risk domains—all the same? Saf Sci 50 (10):1983–1992

Grote G, Künzler C (2000) Diagnosis of safety culture in safety management audits. Saf Sci 34 (1–3):131–150

Hale AR, Swuste P (1998) Safety rules: procedural freedom or action constraint? Saf Sci 29 (3):163–177

Hale AR, Heming BHJ, Smit K, Rodenburg FGTh, van Leeuwen ND (1998) Evaluating safety in the management of maintenance activities in the chemical process industry. Saf Sci 28 (1):21–4445(1–2):305–327

Hale A, Kirwan B, Kjellén U (2007) Safe by design: where are we now? Saf Sci 45:305–327

Hale AR, Guldenmund FW, van Loenhout PLCH, Oh JIH (2010) Evaluating safety management and culture interventions to improve safety: effective intervention strategies. Saf Sci (48) 8:1026–1035

Hébert L (2011) Tools for text and image analysis: an introduction to applied semiotics. Université du Québec à Rimouski, Quebec. http://www.signosemio.com/documents/Louis-Hebert-Tools-for-Texts-and-Images.pdf. Accessed 16 June 2014

Helland A (2009) Dealing with uncertainty and pursuing superior technology options in risk management-the inherency risk analysis. J Hazard Mater 164(2–3):995–1003

Hollnagel E, Woods DD, Leveson N (eds) (2006) Resilience engineering: concepts and precepts. Ashagate Publishing, Aldershot

IEC (1995) Dependability management-risk analysis of technological systems (IEC 300-3-9). Int Electrotech Comm, IEC, Geneva

ILO (2001) ILO-OSH. Guidelines on occupational safety and health management systems. ILO, Geneva

Kapp EA (2012) The influence of supervisor leadership practices and perceived group safety climate on worker safety performance. Saf Sci 50(4):1119–1124

Knegtering B, Pasman HJ (2009) Safety of the process industries in the 21st century: a changing need of process safety management for a changing industry. J Loss Prev Process Ind 22(6):162–168

Le Coze J-C (2013) Outlines of a sensitising model for industrial safety assessment. Saf Sci 51(1):187–201

Leveson NG (2000) Intent specifications: an approach to building human-centered specifications. IEEE Trans Softw Eng 26(1):15–26. http://sunnyday.mit.edu/papers/intent-tse.pdf. Accessed 16 June 2014

Mearns K, Hope L, Ford MT, Tetrick LE (2010) Investment in workforce health: exploring the implications for workforce safety climate and commitment. Accid Anal Prev 42(5):1445–1454

Mohaghegh Z, Mosleh A (2009) Incorporating organizational factors into probabilistic risk assessment of complex socio-technical systems: principles and theoretical foundations. Saf Sci 47(8):1139–1158

Mostia WBL (2009) Got a risk reduction strategy? J Loss Prev Process Ind 22(6):778–782

Myers PM (2013) Layer of protection analysis - quantifying human performance in initiating events and independent protection layers. J Loss Prev Process Ind 26(3):534–546

Neal A, Griffin MA, Hart PM (2000) The impact of organizational climate on safety climate and individual behavior. Saf Sci 34(1):99–109

Neathey F, Sinclair A, Rick J, Ballard J, Hunt W, Denvir A (2006) An evaluation of the five steps to risk assessment, research report 476. Health and Safety Executive, London. http://www.hse.gov.uk/research/rrpdf/rr476.pdf. Accessed 18 June 2014

Olsen E (2010) Exploring the possibility of a common structural model measuring associations between safety climate factors and safety behaviour in health care and the petroleum sectors. Accid Anal Prev 42(5):1507–1516

OHSAS-18002 (2000) Occupational health and safety management systems—guidelines for the implementation of OHSAS 18001. The Occupational Health & Safety Group, Macclesfield, Cheshire

Pasman HJ, Jung S, Prem K, Rogers WJ, Yang X (2009) Is risk analysis a useful tool for improving process safety? J Loss Prev Process Ind 22(6):769–777

Payne SC, Bergman ME, Beus JM, Rodríguez JM, Henning JB (2009) Safety climate: leading or lagging indicator of safety outcomes? J Loss Prev Process Ind 22(6):735–739

Payne SC, Bergman ME, Rodríguez JM, Beus JM, Henning JB (2010) Leading and lagging: process safety climate—incident relationships at one year. J Loss Prev Process Ind 23(6):806–812

Podsakoff PM, MacKenzie SB, Lee JY, Podsakoff NP (2003) Common method biases in behavioral research: a critical review of the literature and recommended remedies. J Appl Psychol 88(5):879–903

Rasmussen J (1997) Risk management in a dynamic society: a modelling problem. Saf Sci 27(2–3):183–213

Reniers GLL, Ale BJM, Dullaert W, Soudank K (2009) Designing continuous safety improvement within chemical industrial areas. Saf Sci 47(5):578–590

Rochlin GI (1999) Safe operation as a social construct. Ergonomics 42(11):1549–1560

Schupp B, Hale A, Pasman H, Lemkovitz S, Goossens L (2006) Design support for the systematic integration of risk reduction into early chemical process. Saf Sci 44(1):37–54

Schöbel M, Manzey D (2011) Subjective theories of organizing and learning from events. Saf Sci 49(1):47–54

Senge PM (1994) The fifth discipline: the art & practice of the learning organization, 2nd edn. Bantam Doubleday/Currency Publishing Group, New York

van Asselt MBA, van Bree L (2011) Uncertainty, precaution and risk governance. J Risk Res 14(4):401–408

van Asselt MBA, Renn O (2011) Risk governance. J Risk Res 14(4):431–449

Väyrynen S, Jounila H, Latva-Ranta J (2014) HSEQ assessment procedure for supplying industrial network: a tool for implementing sustainability and responsible work systems into SMES. In: Ahram T, Karwowsk W, Marek T (eds) Proceedings of the 5th international conference on applied human factors and ergonomics AHFE 2014, Kraków, Poland 19–23 July 2014

Wilson JR (2014) Fundamentals of systems ergonomics/human factors. Special issue: systems ergonomics/human factors. Appl Ergon 45(1):5–13

Wu T-C, Chang S-H, Shu C-M, Chen C-T, Wang C-P (2011) Safety leadership and safety performance in petrochemical industries: the mediating role of safety climate. J Loss Prev Process Ind 24(6):716–721

Zohar D (1980) Safety climate in the industrial organizations: theoretical and applied implications. J Appl Psychol 65(1):96–102

Zohar D (2008) Safety climate and beyond: a multi-level multi-climate framework. Saf Sci 46 (3):376–387

Zohar D, Luria G (2005) A multilevel model of safety climate: cross-level relationships between organization and group-level climates. J Appl Psychol 90(4):616–628

Chapter 12
Leadership Relationships and Occupational Safety and Health Processes in the Finnish Chemical Industry

Toivo Niskanen

Abstract The aim of this study was to explore organizational and technical measures in OSH, and to clarify the potential relationships between legislation, leadership, collaboration, prevention, improvements, monitoring, occupational health care (OHC), training and use of personal protective equipment. The respondents were OSH managers ($N = 85$) and workers' OSH representatives ($N = 120$) from the chemical companies. In the regression analysis, leadership and collaboration displayed a statistically highly significant effect and the quality of legislation had an almost statistically significant effect on continuous improvements. Furthermore, continuous improvements exerted a statistically significant effect and training had a statistically highly significant effect on prevention priorities. In addition, monitoring of the work environment exhibited a highly significant effect, and leadership and collaboration a significant effect on the use of personal protective equipment. The following four factors were found in the collective climate analysis for monitoring of the work environment and risk prevention priorities: Factor 1—Technology and measurements, Factor 2—Management and guidance, Factor 3—Monitoring and Factor 4—Risk management. In the dynamic processes such as present in the chemical industry undertaking a command-and-control approach to implement guidelines to conduct a top-down leadership is inadequate; a fundamentally different OSH system with self-regulation (e.g. Responsible Care Code) is required.

Keywords Leadership · Collaboration · Monitoring · Work environment · Risk · Prevention · Safety · Health · Climate · Chemical industry

---

T. Niskanen (✉)
Ministry of Social Affairs and Health, Occupational Safety and Health Department, Legal Unit, P.O. Box 33, FI-00023 Government, Helsinki, Finland
e-mail: toivo.niskanen@stm.fi

1 Introduction

The focus of this study is to examine the mechanisms through which leadership, collaboration and technical measures can influence OSH issues. Wirth and Sigurdsson (2008) concluded that there is very limited information available about industry's experience with behavioural safety approaches. In the present study, it was hypothesized that leadership and collaboration is the process when OSH leadership broaden and elevate the interests of the workers, generate commitment of individuals to OSH, and when they enable workers to collaborate for the development of OSH: e.g. increased role clarity and increased opportunities for development (applied from Bass et al. 2003). The conclusions of Hale et al. (1997) in their reviews were that there have been few attempts to produce coherent and comprehensive models of safety management systems […] 'There is a need for a framework to represent the complexity and dynamics in this area' (Hale et al. 1997). With respect to safety climate and safety performance, development of leadership and collaboration reflects the ability of the leader to motivate employees to perform beyond their self-interest towards best OSH practices (applied from Kapp 2012). The four practices can be listed as follows: modelling the desired behaviour, inspiring a commitment to the goal, empathizing with employees and intellectually engaging them in the process (applied from Kapp 2012). A better understanding of these experiences would help in identifying not only the best or most promising practices, but also in pinpointing the obstacles and barriers to successful implementations and the knowledge or practical gaps that provide opportunities for research (Wirth and Sigurdsson 2008).

Organizations working in high-risk industries are subject to competing even conflicting demands. These include balancing a concern for OSH with the need for goal attainment, and finding an equilibrium between the need for stability and control with the need for adaptation and change (Colley et al. 2013). The results of Colley et al. (2013) suggested that perceptions of safety climate are sensitive to the relative emphasis that is being placed on these competing values. 'Efforts aimed at increasing the safety of high-risk production systems focus more and more not only on technical and individual-centred measures, but on an integral safety management improving the interplay of technology, organization and human resources' (Grote and Künzler 2000). Furthermore, high-quality collaboration tends to achieve positive OSH outcomes for organizations and workplaces. Although OSH managers may directly affect OSH outcomes at multiple levels (individuals, groups, organizations), often their effects are indirect, mediated through the workers' OSH representative and they gradually appear over substantial periods of time.

Kapp (2012) noted that safety-related behaviour includes a range of activities performed by individuals to maintain a safe workplace and it involves the actions of individuals ensuring that they comply with established safety rules and procedures as well as involving positive changes in worker behaviour. Coyle-Shapiro and Shore (2007) identified three fundamental aspects inherent in the social exchange: relationship, reciprocity and the actual exchange. One fundamental issue raised by

Coyle-Shapiro and Shore (2007) is that the organizational agent in the dyadic relationship often is not defined. They indicated that individuals (in this study—OSH managers and workers' OSH representatives) do not enter into a formal exchange agreement with the organization. Rather, they are dyadic partners with whom there are interactions. As such, managers play an important role in shaping the psychological contract on behalf of the organization (Huy 2002).

Vroom and Yetton (1973) proposed a model of situational leadership, which could be applied to OSH management; this model supports both OSH managers and workers' OSH representatives in deciding the level of worker participation in OSH problem solving. A modification of the situational leadership theory proposed by Hersey and Blanchard (1982) advocates that the amount of direction that a leader provides should be based on the employee's maturity and should be on a continuum consisting of distinctive four styles: autocratic, persuasive, consultative and democratic. The assumption in the situational model of Hersey and Blanchard (1982) can be applied to OSH activities and the sharing of these responsibilities between OSH managers and workers' OSH representatives. The participation of the workers' OSH representatives is a way to promote acceptance of the OSH measures. In addition, the participation of the workers can increase the effectiveness of the OSH measures and improve commitment towards OSH collaboration. Likert (1961) also highlighted the benefits of participative management.

While some OSH managers may have access to effective OSH management systems, this is not invariably the case. It might be more appropriate to focus on how the OSH managers shared actions achieve OSH preventive measures and are these acquired via individual measures rather than overall OSH collaboration between the employer and the workers. It might also be worthwhile focusing on how those different levels of OSH measures can affect self-regulatory processes in the workplaces and how these different interactions can encourage a collective engagement in OSH. The complexity of OSH management can also be appreciated by acknowledging the impact that the OSH context (e.g. national and organizational cultures) may have on leaders, which may in turn, influence on OSH outcomes.

Workplaces are complex systems in which different processes are continually in flux (e.g. OSH management, self-regulation in OSH, collaboration in OSH) as OSH managers with workers' OSH representative and others in the work environment—these interactions results can in very different behavioural OSH outputs. OSH management can also reflect an integration of capacities across roles in creating the workplace culture such as the willingness to undertake cooperation and collaboration between the employer and workers in OSH issues.

OSH management guides the combination of goal orientation in OSH and its regulatory focus. This combination is integrated within the workplaces via self-regulation between OSH managers and workers' OSH representative creating individualized structures that support OSH information processing and behaviour. The OSH managers who appeal to the OSH values of the workers' OSH representatives should be able to project a leadership image which is linked to the organizations' OSH management. The 'trust' theme also emerges in transformational theories (e.g. Bass 1997) and in leader–member exchange theory (Schriesheim et al. 1999, 2000).

The overall goals set for OSH can play an important role in determining the overall OSH behaviour. The goal-setting theory (Locke and Latham 1990, 2002) proposed that individuals make their own estimates about how their decisions will achieve the desired goals. Thus, setting OSH goals can affect an individual's OSH behaviour and their job-related performance (applied from Locke and Latham 2002). Yukl (1999) pointed out that 'the theory could be stronger if the essential influence processes were identified more clearly and used to explain how each type of behaviour can affect each type of mediating variable and outcome'. Furthermore, Yukl (1999) found that if a follower links his or her self-concept with the leader's vision or values, one would expect higher levels of identification.

Companies in the chemical industry have set up the Responsible Care (RC) programme. The underlying concept is that the companies work together by developing global goals in their national associations to continuously improve their health, environmental and safety (HES) performances (ICCA 2014). The significance of the concept has grown in the chemical industry over the last 2 decades, and has been the subject of considerable debate, commentaries and research (Cefic 2014).

In the field of safety climate research, the results of Luria and Yagil (2010) explained the heterogeneity of climate perceptions, and made a strong contribution to safety climate theory. Their basic assumption is that when all members of an organization are exposed to the same environment, they are likely to perceive its components in the same manner. Climates exist for numerous facets of organizational life (Beus et al. 2010). The conceptualization of safety themes according to organizational levels suggests that OSH communications should indicate whether a specific safety issue addresses the individual, the group or the organization (Luria and Yagil 2010). Further, there may be individual difference variables that predict workplace safety behaviour or the degree of safety sensitivity (Beus et al. 2010). There may be different empirical structures embedded into the climate, e.g.: leadership style, training programmes, motivational patterns, participation, innovation, observation and adherence to rules as well as incident and accident reporting systems (Díaz-Cabrera et al. 2007). The general assumption in the model of Olsen (2010) is that top-level management is primarily responsible for defining the correct goals and strategies for the organization and signalling the relative importance of safety in conjunction with other organizational goals. The results of Morrow et al. (2010) indicated that psychological perceptions of work-safety tension are more strongly related to safety behaviour than perceptions of management or co-worker commitment to safety.

Although the research work described above has resulted in a deeper understanding of safety leadership and safety performance, there are still a critical gap exists in the literature. By including the organizational aspects and technical measures of into the present study, it was hoped to clarify the potential relationships between e.g. legislation, leadership, collaboration, prevention, improvements, monitoring, occupational health care (OHC), training and the use of the personal protective equipment.

The Finnish Enforcement Act 44/2006 (Finnish Legislation 2006) stipulates it is the duty of the OSH manager to undertake the necessary measures to organize cooperation between the employer and the employees and to maintain this cooperation in the workplace, as well as contributing to the development of OSH cooperation. In addition, the OSH manager needs to be adequately qualified with regard to the nature of the workplace and the work being done, and the nature of the workplace. Leader–member relationships between OSH managers and workers' OSH representatives share some of the characteristics described by transformational styles of leadership. 'Leadership and Collaboration' is highly reciprocal, with the OSH managers and workers' OSH representatives influencing each other; this type of relationship may be desirable in OSH. The leader–member exchange model is concerned with the hierarchical relationship between a superior and his/her subordinates (Dansereau et al. 1975; Dienesch and Liden 1986).

2 The Research Question and Hypotheses of the Study

The aim of this study was to explore organizational and technical measures in OSH, and clarify the potential relationships between e.g. legislation, leadership, collaboration, prevention, improvements, monitoring, OHC, training and the use of the personal protective equipment. This study explores several research questions: Do the attributes of the effective leadership and collaboration vary in the workplace, and if so, how? What are the relational processes that must be involved in effective collective monitoring of work environment? How are the continuous improvements implemented? How is the on-the-job training in OSH implemented? What is the nature of interactive dynamics in the OSH management? How does the effective leadership and collaboration promote the risk prevention priorities? How and why can relational leadership help to address the challenges of organizing in effects of OHC services?

Based on the results of other researchers, this study advances four hypotheses:

The background to Hypothesis 1: 'Leadership and Collaboration' will be effective if the leader's competence matches the OSH complexity and OSH improvement required by the work (here it refers to the processes in the chemical industry). Previously, Piccolo and Colquitt (2006) described a mechanism for explaining the effects of 'Leadership and Collaboration' rooted in the job. Carmeli et al. (2013) found that the OSH collaboration is a necessary relational mechanism which creates the space to allow more positive interactions between and workers. In this context, management team members discuss organizational issues and create the potential to effectively gather and process relevant OSH information for the OSH collaboration. Zohar and Tenne-Gazit (2008) highlighted the effects of transformational leadership on the safety climate.

Grote (2012) concluded that in recent years, there has been a tendency to move away from prescriptive regulation which specifies in great detail how an organization has to carry out its operations, towards goal-oriented legislation, which promotes self-regulation. Attention needs to be paid to how inputs in implementing the legislation and self-regulation can be combined to create effective OSH outputs and how these relationships should be integrated. When the external regulation and self-regulation coexist, these two perspectives need to be aligned by assigning responsibility in appropriate ways that match the coordination capabilities of the respective actors (Grote 2012).

One of the more powerful influences a leadership can have on followers is in the 'management of meaning', as leaders define and shape the 'reality' in which their followers work (Piccolo and Colquitt 2006). Uhl-Bien et al. (2012, p. 310) have considered relational leadership as not simply leader–member exchange (and its more limited set of measures) but as a wider examination of leadership relationships (dyadic and collective) and relational processes (e.g. practices, improvements) in organizations that contribute to the generation of leadership. The focus of the work of Lord and Din (2012, p. 37) was on events, i.e. there is a tendency for the experience of different events coalesce to produce behavioural improvements. They also examined leadership skills and how leaders can influence the ways in which events occur to produce performance outcomes in others. When leadership processes affect some individuals independently of other group members, one might expect leadership to produce differential compositional effects on group characteristics or processes (Lord and Din 2012, p. 49). Dyadic-level leadership processes (i.e. leader–member exchanges) could also be included within this category described by Lord and Din (2012, p. 49), although they also may be influenced by group and organizational attributes.

Hypothesis 1 The aggregated variables 'Leadership and Collaboration' and 'Quality of Legislation' have a positive effect on the aggregated variable 'Continuous Improvements'.

The background to Hypothesis 2: Reason (1997) argued that an informed culture has four characteristics: (1) a reporting culture, (2) a just culture, (3) a flexible culture and (4) a learning culture. Risk assessment and continuous Improvements should represent an integral part of a company's safety management system. The objectives of the OSH Act (Finnish Legislation 2002) are to improve the working conditions in order to ensure the working capacity of employees as well as to prevent occupational accidents and diseases, and eliminate other hazards. Grote (2012) stated that the move towards goal-oriented regulation could also be understood in the context of a general trend towards acknowledging the need to cope with uncertainty instead of trying to manage it away. In behavioural safety, management participation plays a key role in supporting the process by providing feedback and pinpointing critical phases, contributing to the development of checklists, providing performance feedback and reinforcement directly to workers and safety committees (Wirth and Sigurdsson 2008). In view of the increasing complexity of socio-technical systems, the possibility of anticipation and perfect prediction being the basis for legal regulation becomes less and less feasible. Instead, resilience through prompt adaptation to emerging threats is required which relies on decentralized small-scale trial and error, which can be staged and monitored by the regulatory agencies (Grote 2012).

Since leaders are important role models and should be aware of the critical aspects of the situation, they can directly influence how experiences gained during different events can be utilized in the future (Zohar and Tenne-Gazit 2008). Leaders who are able to create positive and supportive environments may increase the likelihood that OSH managers and workers' OSH representative will cooperate in complex tasks or to learn from their experiences and wisely use their resources. This is because this kind of organizational climate allows these individuals to effectively promote OSH measures. Leaders can also affect how others utilize information through their communication of its impact on the commitment (applied from Lord and Din 2012, p. 47) to promoting OSH to avoid exposures for dangerous chemicals.

Hypothesis 2 The aggregated variables 'Occupational Health Care (OHC)' and 'Leadership and Collaboration' will show a positive effect on the aggregated variable 'Monitoring of Work Environment'.

The background to Hypothesis 3: The OHC (Finnish Legislation 2001) means the activities of OHC professionals and experts that the employer has a duty to arrange by law and which are used to promote the prevention of work-related illnesses and accidents, the healthiness and safety of the work and the working environment. An organization creates and maintains OSH through implementation and monitoring of all its leadership activities (e.g. on-the-job training, risk management, management of change, auditing, inspection and learning from experience) (Le Coze 2013). This involves an extensive feedback–feedforward loop based on a designed interrelationship between specific OSH activities or functions (Le Coze 2013). Furthermore, safety is the outcome of the quality of this feedback–feedforward loop within performance and improvements. In the USA, OSHA (2012, p. 4) states that the workers have the right to know about chemical hazards. In addition, the employers must: (1) Inform employees about hazards through training, labels, alarms, colour-coded systems, chemical information sheets and other methods; (2) Train employees in a language and vocabulary they can understand (OSHA 2012, p. 4).

Instead of having to continuously react to changes in external demands and adapt their OSH improvements accordingly, companies can proactively manage safety in a way that best fits their specific situation (Grote 2012). In order to employ rules in ways that support the overall objective of adequately balancing stability and flexibility, the nature of the rules themselves needs to be examined. Leaders can provide feedback that facilitates training and learning in a way that emphasizes comparisons among individuals, thereby making performance goal orientation more relevant; or they can focus their feedback on task skills, placing more emphasis on learning goal orientations (Lord and Din 2012, p. 39).

Hypothesis 3 The aggregated variables 'Continuous Improvements' and 'On-the-job Training' display a positive effect on the aggregated variable 'Risk Prevention Priorities'.

The background to Hypothesis 4: Leadership can be viewed as the process of influencing others to understand and agree about what needs to be done and how to achieve the goal as well as the process of facilitating individual and collective efforts to accomplish shared objectives (Yukl and Becker 2006; Yukl 1999). In the USA, OSHA (2012, p. 8) stipulates that in some situations, it is not possible to completely eliminate a hazard or reduce exposures to a safe level, so personal protective equipment must often be used by workers or be used in addition to other hazard control measures. Furthermore, employers are responsible for knowing when protective equipment will be needed.

Constructionism assumes that social reality is not separate from an individual's sense of reality since both are intimately interwoven and shaped by each other in everyday interactions (Cunliffe 2008). Constructionist theorists approach leadership not as a phenomenon embodied in individuals but as an organizing process with a foundation in task accomplishment (Fairhurst and Grant 2010). These investigators consider the actual behaviours and interactions of individuals as part of a broader organizational process, where patterned interactions and networks of relationships contribute to define the ultimate outcomes (Uhl-Bien et al. 2012, p. 307).

Hypothesis 4 The aggregated variables 'Monitoring of Work Environment' and 'Leadership and Collaboration' will exhibit a positive effect on the aggregated variable 'Use of the Personal Protective Equipment'.

3 Data Collection, Methods and Participants

3.1 Research Materials

The online questionnaire research was conducted in September–October 2011. The questionnaire survey was implemented by using a data collection programme on the Internet. The questionnaire form was filled in by OSH managers and by workers' OSH representatives. The online electronic questionnaire form was filled in by OSH managers and by workers' OSH representatives of the Finnish chemical industry's workplaces. The employer respondents were selected from the register of OSH managers (including their e-mail address) kept by the Centre of Occupational Safety. The questionnaire was sent to 357 OSH managers, of whom 85 responded, resulting in a response rate of 24 %. The employee respondents were OSH representatives in different companies, which were selected from the registers of the Finnish Industrial Union TEAM (blue-collar workers) and of the Finnish Trade Union PRO (clerical workers). The questionnaire was sent to 480 workers' OSH representatives, of whom 120 (25 %) responded.

The participants were informed about the purpose, aims and methods of the study and assured that participation was voluntary. In addition, they were told that the anonymity of the participants would be assured during and after the study, and data security would be guaranteed. They were also given the opportunity to contact the researcher to obtain more information about the study.

In this analysis, the data of OHS managers and workers' OSH representatives are combined together in order to execute the statistical analysis in a more reliable manner.

3.2 Methods and Statistical Analysis

OSH managers and workers' OSH representatives were asked to evaluate how well the different OSH measures were accomplished by recording their responses on a Likert scale ranging from 1 to 5 as follows: strongly disagree, somewhat disagree, somewhat agree and strongly agree. Statistical analysis was done with SAS software (2005). In the factor analysis, the varimax rotation was used. In this study, Hypotheses 1–4 were tested using regression analysis. The Kaiser–Meyer–Olkin (KMO) provides a measure of sampling adequacy as a diagnostic measure in the factor analysis. The following labels are given to values of KMO (Kaiser 1974): 0.60–0.69 'mediocre', 0.70–0.79 'middling', 0.80–0.89 'meritorious' and 0.90–1.00 'marvelous'.

4 Results of the Questionnaire Study

4.1 Testing of Hypothesis 1 in OHS Managers and Workers' OSH Representatives

In this analysis, the data of OHS managers ($N = 85$) and workers' OSH representatives ($N = 120$) are combined together.

The correlation coefficients of all aggregated variables for 'Continuous Improvements' (Question 1, Appendix 1), 'Leadership and Collaboration' (Question 2) and 'Quality of Legislation' (Question 3) were statistically highly significant ($p < 0.001$) (Table 1).

In the regression model (Table 2), the effects of 'Leadership and Collaboration' (independent variable) and 'Quality of Legislation' (independent variable) on 'Continuous Improvements' (dependent variable) confirmed the goodness of fit (R-squared = 0.38). The statistical significance of the model was highly significant as assessed by F-test ($F_{1,196} = 53.81$, $p < 0.0001$).

The fit of individual parameters was tested by t-test (Table 2). In the regression analysis, 'Leadership and Collaboration' exhibited a statistically highly significant ($p < 0.001$) effect and 'Quality of Legislation' a statistically nearly significant ($p < 0.05$) effect on 'Continuous Improvements'.

Table 1 Pearson correlation coefficient of all aggregated variables of 'Continuous Improvements', 'Leadership and Collaboration' and 'Quality of Legislation'

Variable	N	Mean	Standard deviation	Number of items in scale	Q1. 'Continuous Improvements'	Q2. 'Leadership and Collaboration'	Q3. 'Quality of Legislation'
Q1. 'Continuous Improvements'	206	3.27	0.68	7	1.000		
Q2. 'Leadership and Collaboration'	205	3.11	0.54	9	0.6022***	1.000	
Q3. 'Quality of Legislation'	197	2.74	0.61	9	0.3128***	0.2973***	1.000

***$p < 0.001$

Table 2 R-square and the parameter estimates in the regression analysis with respect to the relationships of the aggregated variable 'inputs' (independent) and 'outputs' (dependent—'Continuous Improvements')

Source	df	Sum of squares	Mean square	F value	Prob. level	R-square
Model	2	1217.01	608.50	58.31	<0.0001	0.38
Error	194	2024.53	10.43574			
Corrected total	196	3241.54				

Independent variable	Parameter estimate	Standard error	t-value	Prob. level	95 % confidence limit	
Intercept	6.93	1.53	4.54	<0.0001***	3.92	9.95
Q2. 'Leadership and Collaboration'	0.47	0.05	9.29	<0.0001***	0.37	0.57
Q3. 'Quality of Legislation'	0.11	0.04	2.50	0.0131*	0.02	0.20

*** $p < 0.001$
* $p < 0.05$

4.2 Testing of Hypothesis 2 in OHS Managers and Workers' OSH Representatives

In this analysis, the data of OHS managers ($N = 85$) and workers' OSH representatives ($N = 120$) are combined together.

The correlation coefficients of all aggregated variables for the 'Monitoring of Work Environment' (Question 4), 'Occupational Health Care' (Question 5) and 'Leadership and Collaboration' (Question 2) were statistically highly significant ($p < 0.001$) (Table 3).

In the regression model (Table 4), the effects of 'Occupational Health Care' (independent variable) and 'Leadership and Collaboration' (independent variable) on 'Monitoring of Work Environment' (dependent variable) confirmed the goodness of fit (R-squared = 0.55). The statistical significance of the model was highly significant as assessed by F-test ($F_{1,196} = 119.47$, $p < 0.0001$).

The fit of individual parameters was tested by t-test (Table 4). In the regression analysis, both 'Occupational Health Care' and 'Leadership and Collaboration' exhibited a statistically highly significant ($p < 0.001$) effect on 'Monitoring of Work Environment'.

4.3 Testing of Hypothesis 3 Among OHS Managers and Workers' OSH Representatives

In this analysis, the data of OHS managers ($N = 85$) and workers' OSH representatives ($N = 120$) are combined together.

The correlation coefficients of all aggregated variables for 'Risk Prevention Priorities' (Question 6), 'Continuous Improvements' (Question 1) and 'On-the-job Training' (Question 7) were statistically highly significant ($p < 0.001$) (Table 5).

In the regression model (Table 6), the effects of 'Continuous Improvements' (independent variable) and 'On-the-job Training' (independent variable) on 'Risk Prevention Priorities' (dependent variable) confirmed the goodness of fit (R-squared = 0.55). The statistical significance of the model was highly significant as assessed by F-test ($F_{1,196} = 119.47$, $p < 0.0001$).

The fit of individual parameters was tested by t-test (Table 6). In the regression analysis, 'Continuous Improvements' exhibited a statistically significant ($p < 0.01$) effect and 'On-the-job Training' a statistically highly significant ($p < 0.001$) effect on 'Risk Prevention Priorities'.

Table 3 Pearson correlation coefficient of all aggregated variables of 'Monitoring of Work Environment', 'Content of OHC' and 'Leadership and Collaboration'

Variable	N	Mean	Standard deviation	Number of items in scale	Q4. 'Monitoring of work environment'	Q5. 'Occupational Health Care'	Q2. 'Leadership and Collaboration'
Q4. 'Monitoring of Work Environment'	203	3.08	0.86	7	1.000		
Q5. 'Occupational Health Care'	202	3.26	0.69	8	0.7115***	1.000	
Q2. 'Leadership and Collaboration'	205	3.11	0.54	9	0.6391***	0.6894***	1.000

***$p < 0.001$

Table 4 R-square and the parameter estimates in the regression analysis with respect to the relationships of the aggregated variable 'inputs' (independent) and 'outputs' (dependent—'Monitoring of Work Environment')

Source	df	Sum of squares	Mean square	F value	Prob. level	R-square
Model	2	2968.37	1484.18	119.47	<0.0001	0.55
Error	198	2459.84	12.42			
Corrected total	200	5428.21				

Independent variable	Parameter estimate	Standard error	t-value	Prob. level	95 % confidence limit	
Intercept	0.01	1.51	0.01	0.9974	−2.97	2.98
Q5. 'Occupational Health Care'	0.49	0.06	7.86	<0.0001***	0.37	0.61
Q2. 'Leadership and Collaboration'	0.30	0.07	4.21	<0.0001***	0.16	0.45

***$p < 0.001$

Table 5 Pearson correlation coefficient of all aggregated variables of 'Risk Prevention Priorities', 'Continuous Improvements' and 'On-the-job Training'

Variable	N	Mean	Standard deviation	Number of items in scale	Q6. 'Risk Prevention Priorities'	Q1. 'Continuous Improvements'	Q7. 'On-the-job Training'
Q6. 'Risk Prevention Priorities'	205	3.02	0.72	9	1.000		
Q1. 'Continuous Improvements'	206	3.27	0.58	7	0.562***	1.000	
Q7. 'On-the-job Training'	206	3.16	0.65	7	0.643***	0.725***	1.000

*** $p < 0.001$

Table 6 R-square and the parameter estimates in the regression analysis with respect to the relationships of the aggregated variable 'inputs' (independent) and 'outputs' (dependent—'Risk Prevention Priorities')

Source	df	Sum of squares	Mean square	F value	Prob. level	R-square
Model	2	3798.84	1899.42	77.05	<0.0001	0.43
Error	202	4979.72	24.65			
Corrected total	204	8778.56				

Independent variable	Parameter estimate	Standard error	t-value	Prob. level	95 % confidence limit	
Intercept	4.00	2.02	1.98	0.0491*	0.02	7.99
Q1. 'Continuous Improvements'	0.33	0.12	2.65	0.0087**	0.08	0.57
Q7. 'On-the-job Training'	0.71	0.11	6.46	<0.0001***	0.49	0.93

***$p < 0.001$
**$p < 0.01$
*$p < 0.05$

4.4 Testing of the Hypothesis 4 Among OHS Managers and Workers' OSH Representatives

In this analysis, the data of OHS managers ($N = 85$) and workers' OSH representatives ($N = 120$) are combined together.

The correlation coefficients of all aggregated variables for 'Use of the Personal Protective Equipment' (Question 8), 'Monitoring of Work Environment' (Question 4) and 'Leadership and Collaboration' (Question 2) were statistically highly significant ($p < 0.001$) (Table 7).

In the regression model (Table 8), the effects of 'Monitoring of Work Environment' (independent variable) and 'Leadership and Collaboration' (independent variable) on 'Use of the Personal Protective Equipment' (dependent variable) confirmed the goodness of fit (R-squared $= 0.45$). The statistical significance of the model was highly significant as assessed by F-test ($F_{1,204} = 80.17$, $p < 0.0001$).

The fit of individual parameters was tested by t-test (Table 8), which indicated a statistically highly significant ($p < 0.001$) effect of 'Monitoring of Work Environment' and a statistically nearly significant ($p < 0.01$) effect of 'Leadership and Collaboration' on 'Use of the Personal Protective Equipment'.

4.5 Hypothesis Verification

This study verified the four hypotheses H1–H4.

H1 In the regression analysis, 'Leadership and Collaboration' exhibited a statistically highly significant ($p < 0.001$) effect and 'Quality of Legislation' showed a statistically nearly significant ($p < 0.05$) effect on 'Continuous Improvements'

H2 In the regression analysis, both 'Occupational Health Care' and 'Leadership and Collaboration' exhibited a statistically highly significant ($p < 0.001$) effect on 'Monitoring of Work Environment'

H3 In the regression analysis, 'Continuous Improvements' exhibited a statistically significant ($p < 0.01$) effect and 'On-the-job Training' a statistically highly significant ($p < 0.001$) effect on 'Task Performance Risk Prevention Priorities'

H4 In the regression analysis, 'Monitoring of Work Environment' exhibited a statistically highly significant ($p < 0.001$) effect and 'Leadership and Collaboration' a statistically significant ($p < 0.01$) effect on 'Use of the Personal Protective Equipment'.

Table 7 Pearson correlation coefficient of all aggregated variables of workers for the followings: 'Use of the Personal Protective Equipment', 'Monitoring of Work Environment' and 'Leadership and Collaboration'

Variable	N	Mean	Standard deviation	Number of items in scale	Q8. 'Use of the Personal Protective Equipment'	Q4. 'Monitoring of work environment'	Q2. 'Leadership and Collaboration'
Q8. 'Use of the Personal Protective Equipment'	206	3.14	0.77	7	1.000		
Q4. 'Monitoring of work environment'	205	3.01	0.80	7	0.659***	1.000	
Q2. 'Leadership and Collaboration'	203	3.24	0.72	8	0.550***	0.716***	1.000

***$p < 0.001$

Table 8 R-square and the parameter estimates in the regression analysis with respect to the relationships of the aggregated variable 'inputs' (independent) and 'outputs' (dependent—'Use of the Personal Protective Equipment')

Source	df	Sum of squares	Mean square	F value	Prob. level	R-square
Model	2	2623.47	1311.74	80.17	<0.0001	0.45
Error	200	3272.56	16.36			
Corrected total	202	5896.03				

Independent variable	Parameter estimate	Standard error	t-value	Prob. level	95 % confidence limit	
Intercept	7.11	1.33	5.36	<.0001***	4.50	9.73
Q4. 'Monitoring of work environment'	0.52	0.07	7.15	<.0001***	0.38	0.66
Q2. 'Leadership and Collaboration'	0.15	0.07	2.17	0.0308*	0.01	0.29

*** $p < 0.001$
* $p < 0.05$

4.6 Collective Climate

4.6.1 Climate for 'Monitoring of Work Environment' and 'Risk Prevention Priorities'

The factor loadings of the responses to the item statements about how the OSH managers and workers OSH representatives assess 'Monitoring of Work Environment' (Question 4) and 'Risk Prevention Priorities' (Question 6) when calculated with varimax rotation (Table 9).

Standardized Cronbach coefficient alpha is 0.91. Kaiser–Meyer–Olkin measure of sampling adequacy is 0.84.

Factor 1 can be named as 'Technology and Measurements of Exposure', Factor 2 a s 'Management and Guidance', Factor 3 as 'Monitoring' and Factors 4 as 'Risk Management'. Weighted variances explained by each factor are as follows: Factor 1—18 %, Factor 2—15 %, Factor 3—14 %, Factor 4—12 %.

4.6.2 Climate for 'Continuous Improvements' and 'Leadership and Collaboration'

The factor loadings of the responses to the item statements about how the OSH managers and workers OSH representatives about 'Continuous Improvements' (Question 1) and 'Leadership and Collaboration' (Question 2) when calculated with varimax rotation (Table 10).

Standardized Cronbach coefficient alpha is 0.91. Kaiser–Meyer–Olkin measure of sampling adequacy is 0.84.

Factor 1 can be named as 'Leadership and training', Factor 2 as 'Participation of workers', Factor 3 as 'Best practices and legislation'. Weighted variances explained by each factor are as follows: Factor 1—17 %, Factor 2—16 %, Factor 3—13 %.

4.6.3 Climate for 'OSH Enforcement Operations', 'Effects of the Current OSH Inspection' and 'Enforced Policy of the Current OSH Inspection'

The factor loadings of the responses to the item statements about how the OSH managers and workers OSH representatives about 'OSH enforcement operations', 'Effects of the current OSH inspection' and 'Enforced Policy of the Current OSH Inspection' (When they had participated in an OSH inspection) when calculated with varimax rotation (Table 11).

Standardized Cronbach coefficient alpha is 0.63. Kaiser–Meyer–Olkin measure of sampling adequacy is 0.74.

Factor 1 is named as 'Strength of OSH enforcement', Factor 2 as 'Effects on OSH systems', Factor 3 as 'Expertise within OSH inspections', Factors 4 as

Table 9 The item statements about 'Monitoring of Work Environment' (Question 4) and 'Risk Prevention Priorities' (Question 6) with their rotated factor loadings and communality

Statements	Factor 1 loading	Factor 2 loading	Factor 3 loading	Factor 4 loading	Communality	Weight
1. Local exhaust ventilation is used in work stations when necessary	**0.74**	0.13	0.10	0.23	0.62	2.66
2. The effectiveness of ventilation is followed up and the ventilation equipment is serviced regularly	**0.73**	0.19	0.14	0.24	0.65	2.84
3. There is a general air conditioning system that functions well, and the amount of both exhausted air and replacement air are large enough	**0.71**	0.10	0.07	0.27	0.58	2.40
4. Any exposure of employees to chemicals is followed up with the help of biological monitoring carried out by the OHC service provider (urine and blood tests)	**0.51**	0.23	0.15	−0.04	0.34	1.52
5. It is ensured that occupational exposure level limit (OEL) values known to be harmful are not exceeded	**0.49**	0.31	0.24	0.30	0.49	1.95
6. Work air impurities are followed up through regular measurements of occupational exposure limit (OEL) values	**0.46**	0.38	0.20	0.17	0.42	1.72
7. The activities of the management of the enterprise represent the best way to tangibly promote the monitoring of the work environment	0.17	**0.80**	0.28	0.35	0.87	7.99
8. The management of the enterprise represents the best way to tangibly promote the safe handling of chemicals	0.30	**0.71**	0.15	0.36	0.74	3.81
9. The cooperation (between the employer and employees) on OSH matters represents the best way to tangibly promote the safe handling of chemicals	0.33	**0.48**	0.43	0.13	0.54	2.18

(continued)

Table 9 (continued)

Statements	Factor 1 loading	Factor 2 loading	Factor 3 loading	Factor 4 loading	Communality	Weight
10. The training has been arranged to secure that plans and guidance work in practice	0.36	**0.43**	0.24	0.10	0.37	1.59
11. The requirements of the OSH legislation represent the best way to tangibly promote the monitoring of the work environment	0.06	0.24	**0.93**	0.18	0.95	21.99
12. The activities of the management of the enterprise represent the best way to tangibly promote the monitoring of the work environment	0.26	0.13	**0.73**	0.23	0.67	3.04
13. The cooperation (between the employer and employees) on OSH matters represents the best way to tangibly promote the monitoring of the work environment in practice	0.21	**0.49**	**0.58**	−0.05	0.61	2.61
14. The employer has chosen such chemicals for use that cause least OSH harm	0.12	0.22	0.14	**0.74**	0.63	2.71
15. The employer has chosen such production methods that cause least OSH harm	0.26	0.08	0.07	**0.70**	0.58	2.36
16. My workplace is actively searching for safer alternatives to replace dangerous chemicals	0.25	0.33	0.22	**0.51**	0.49	1.95

Table 10 The item statements about 'Continuous Improvements' (Question 1) and 'Leadership and Collaboration' (Question 2) with their rotated factor loadings and communality

Statements	Factor 1 loading	Factor 2 loading	Factor 3 loading	Communality	Weight
1. The OSH manager actively promotes the OSH activities at our workplace	**0.78**	0.25	0.16	0.70	3.28
2. HES (health, environment and safety) management actively promotes the OSH activities at our workplace	**0.68**	0.32	0.17	0.59	2.45
3. We solve and take care of OSH issues together	**0.57**	0.36	0.37	0.60	2.47
4. The OHC service provider is active in OSH matters at my workplace	**0.44**	0.25	0.25	0.31	1.45
5. The activities of the management of the enterprise represent the best way to tangibly promote the Continuous Improvements of safety in practice	**0.40**	0.34	0.34	0.39	1.64
6. Only a few persons take care of OSH matters at my workplace	−0.42	−0.20	−0.13	0.24	1.51
7. The OSH representative receives all necessary information on OSH matters (accidents occurred, the workplace survey drawn up by the OHC service provider, etc.)	0.24	**0.92**	0.12	0.91	11.83
8. The personnel is informed about matters relating to OSH	0.28	**0.71**	0.23	0.63	2.72
9. The OSH representative receives appropriate OSH training	0.26	**0.60**	0.10	0.44	1.80
10. OSH training has been utilized in the OSH activities at my workplace	**0.50**	**0.53**	0.20	0.57	2.33
11. Pregnant employees are taken into account in the safety practices	−0.02	0.14	**0.70**	0.51	2.05
12. The cooperation (between the employer and employees) on OSH matters represents the best way to tangibly promote the Continuous Improvements of safety in practice	0.22	0.16	**0.62**	0.46	1.86
13. The requirements of the OSH legislation represent the best way to tangibly promote the Continuous Improvements of safety in practice	0.11	0.06	**0.47**	0.24	1.31
14. The safety practices at my workplace are also applied to subcontractors working in the enterprise	0.29	0.19	**0.43**	0.31	1.44
15. Occurred deviations from safety are investigated, for example near accidents, accidents, etc.	0.18	0.07	0.39	0.19	1.24
16. The safety aspects have been taken into account in sufficient detail in the instructions for work in order to make the working safe	0.33	0.02	0.38	0.26	1.35

Table 11 The item statements about 'OSH enforcement operations' (Question 9), 'Effects of the current OSH inspection' (Question 10) and 'Enforced Policy of the Current OSH Inspection' (Question 11) with their rotated factor loadings and communality

Statements	Factor 1 loading	Factor 2 loading	Factor 3 loading	Factor 4 loading	Factor 5 loading	Communality	Weight
1. The OSH inspector should more often impose binding obligations to remedy deficiencies at workplaces	**0.87**	0.08	−0.10	0.22	−0.04	0.83	5.89
2. The OSH inspector should strengthen the enforcement of certain sections of relevant OSH legislation	**0.75**	0.14	−0.24	0.09	0.06	0.65	2.88
3. The OSH inspector should visit my workplace more often	**0.75**	−0.02	−0.09	−0.01	**0.59**	0.92	11.78
4. OSH inspectors' visits should be made more often to our workplace	**0.71**	0.01	−0.08	0.04	**0.54**	0.81	5.25
5. The OSH inspector should follow up that OSH management systems are being applied	**0.63**	0.01	−0.10	0.21	0.22	0.50	1.99
6. The OSH inspector should improve the quality of his/her inspections	0.36	−0.33	−0.23	0.12	0.13	0.32	1.48
7. The inspection led to a more systematic development of OSH	0.01	**0.82**	0.12	0.01	−0.02	0.69	3.19
8. The inspection triggered the preparation of some OSH documents, or to a rewriting some existing documents in a more detailed manner	0.13	**0.76**	−0.08	−0.15	0.00	0.62	2.64
9. The inspection triggered the correction of the deficiencies observed	−0.05	**0.69**	0.24	−0.00	0.04	0.53	2.14
10. During the inspection, we received new information on the legal obligations concerning our workplace	−0.01	**0.67**	0.12	−0.05	0.09	0.47	1.90
11. The inspection was carried out in a professional manner	0.05	**0.45**	**0.40**	−0.09	0.12	0.39	1.65
12. During the inspection, the inspector obtained a truthful picture of our workplace	−0.17	0.25	**0.93**	−0.01	−0.21	0.99	11.99
	−0.38	0.01	**0.46**	−0.02	−0.15	0.38	1.62

(continued)

Table 11 (continued)

Statements	Factor 1 loading	Factor 2 loading	Factor 3 loading	Factor 4 loading	Factor 5 loading	Communality	Weight
13. During the inspection, the employees working at their workstations could express their opinions freely							
14. All severe hazards and risks in our workplace were not evaluated during the inspection	0.25	−0.14	**−0.40**	0.11	0.17	0.28	1.38
15. The OSH inspector should give more advice to workplaces in order to help them to exceed the minimum level laid down by law	0.15	−0.12	−0.13	**0.97**	0.08	0.99	12.01
16. The OSH inspector should give more advice	0.33	−0.06	0.01	**0.55**	0.13	0.43	1.76
17. During the inspection, the kinds of deficiencies or faults were assessed that would not have been tackled without the inspection	0.11	0.38	−0.14	0.07	**0.42**	0.35	1.54
18. The OSH enforcement should be restricted only to ensuring that the minimum level laid down by law is obeyed	−0.05	0.08	0.09	−0.17	−0.36	0.17	1.21
19. Too much time was used for the inspection	−0.19	−0.21	0.09	0.03	**−0.49**	0.33	1.50

'Advices from OSH inspectors' and Factor 5 as 'Quantity and quality of OSH inspections'. Weighted variances explained by each factor are as follows: Factor 1—18 %, Factor 2—15 %, Factor 3—9 %, Factor 4—8 % and Factor 5—7 % .

5 Discussion

5.1 Theoretical and Practical Implication

This present results support the conclusions of Carmeli et al. (2013) about 'Leadership and Collaboration'; these investigators found that senior managers need to seek ways to develop a working and resilient collaboration that can build and nurture coping mechanisms and enable continuous adaptations. Furthermore, Carmeli et al. (2013) suggested that top management team members should pay closer attention to the relational connection within a team because it affects the type of decision-making process in which they engage. In addition, the mode of decision-making process may affect the quality of their decisions and responses, which ultimately determines the viability and functionality of the OSH system as a whole.

This study supports the conclusions of Dienesch and Liden (1986) in terms of their structured theoretical basis for leader–member exchange, and it indicates that a multidimensional framework for the construct can be applied in exploring OSH collaboration and performance concerned with how OSH managers (as leaders) develop relationships with workers' OSH representatives (as followers). For example, this can offer a valuable framework for studying the best ways to achieve OSH collaboration and Task Performance in organizational relationships. The results of Hale and Borys (2014b) emphasize the importance of targeting the dialogue between workforce and line management; they stated that the most crucial factor is in ensuring that an organization learns and changes.

The present results support the conclusions of Hale and Borys (2014a) in terms of 'Leadership and Collaboration' that there is a need to view rule sets as dynamic and to place the focus of their management on the processes around monitoring and change (flexibility) and not only on development (in this study—'Risk Prevention Priorities') and communication. These characteristics define the gap between procedures and practice (Hale and Borys 2014a). Consequently, in its evaluation of 'Risk Prevention Priorities' this study supports the results of Grote and Künzler (2000) that while dealing with socio-technical systems, the technical and social subsystems of a work system need to be jointly optimized to allow maximum efficiency in the implementation of the system's primary task.

These results support the conclusions of Dienesch and Liden (1986) about the nature of the leader–member exchange. The similar responses provided by OSH managers (as leaders) and workers' OSH representatives (as followers) in the context of the present study are evidence that the quality of the leader–member exchange develops over time. When OSH managers and workers' OSH representatives cooperate, then this collaboration leads to them making similar evaluations in their responses. This is the kind of dyadic relationship which is dependent upon the degree to which each member of the dyad fulfils their expected work roles.

The trust in the hierarchical leader is directly correlated to the extent of shared leadership in organization groups, and it serves as a facilitating force to ensure smooth social interactions, which in turn directly affect the group's ability to share leadership effectively (Wassenaar and Pearce (2012, p. 368)). Moreover, the components of the leader–member exchange relationship are themselves rooted in the OSH exchange relationship between OSH managers and workers' OSH representatives. The dimensions of mutual contribution reflect the concept proposed by Dienesch and Liden (1986) who labelled it as 'mutuality' in an effort to emphasize the reciprocal nature of the relationship of OSH managers and workers' OSH representatives. The present study found evidence of leader–member exchange, dyadic relationships and performance excellence which will be described in the following paragraphs.

In theory, this study support the conclusions of Ayman and Adams (2012, p. 233) who used a systems approach to examine group relationships—namely the input–process–output (I-P-O) model as a heuristic basis for conceptualizing several contextual factors. This can be applied in the evaluation of the processes of OSH effectiveness as examined in the following topics: 'Leadership and Collaboration', 'Monitoring of Work Environment', 'Continuous Improvements', 'On-the-job Training', 'Risk Prevention Priorities' and 'Use of the Personal Protective Equipment'. This makes it possible to achieve a dynamic process perspective, permitting examination of the reciprocal effects between inputs and processes and the relationships between processes and outputs.

In its assessments of 'Monitoring of Work Environment' and in 'Continuous Improvements', this study supports the conclusions of Uhl-Bien et al. (2012, p. 291), i.e. when leaders (OSH managers) and subordinates (workers' OSH representatives) enjoy good and supportive collaboration, they report more positive attitudes (in this study—'Monitoring of Work Environment') and behavioural outcomes (in this study—'Continuous Improvements'). As a consequence, the workplace and leadership dynamics become more effective. Inclusive leadership emphasizes the importance of recognizing the relational context in an interpersonal process in which leaders (employers' OSH representatives) provide different resources 'Continuous Improvements' directed towards goal attainment in OSH, and followers (workers' OSH representatives) determine whether leaders are actually accorded legitimacy to lead through the 'Leadership and Collaboration'. In this way, followers can be active participants in the OSH process, e.g. in they are described here as 'Risk Prevention Priorities'. This is essential for attaining organizational OSH goals. Thus, with respect to its findings about 'Leadership and Collaboration', this study supports the conclusions of Uhl-Bien et al. (2012, p. 294) that leaders engage in an inclusive process by building and bolstering leadership practices through climates that encourage trust.

Rules are usually meant to create particular routines and best practices (in this study—'On-the-job Training') and also convey the reasoning behind the required behaviour (in this study—'Use of the Personal Protective Equipment') in the sense of a routine in principle (Grote 2012). The findings of the section of 'Monitoring of Work Environment' examined here support the conclusions of Reniers (2009) that a very important aspect of effective management of change systems (in this study —'Monitoring of Work Environment' and 'Leadership and Collaboration') is that there are procedures in place that ensure that there is continued safe and reliable operation (in this study—'Risk Prevention Priorities') of complex facilities.

Consequently, instead of having to continuously react to changes in external demands and adapt their OSH management accordingly, companies can proactively manage safety in a way that best fits their specific situation (Grote 2012). In order to employ rules in ways that support the overall objective of adequately balancing stability and flexibility, the nature of the rules themselves needs to be examined. Furthermore, the workers claimed that many workplace exposures can be traced to actions taken which reflect the commitment of and cooperation with line management. This study supports the findings of Kapp (2012) that transformational leadership practices on the part of the front-line supervisor improve the safety-related behaviour of subordinates, including greater compliance with safety procedures, increased 'Use of the Personal Protective Equipment'. This study supports the conclusions of Wirth and Sigurdsson (2008) about the behavioural safety (in this study—e.g. 'Use of the Personal Protective Equipment'). This can actually be conceptualized as fitting into any level of the hierarchy whenever implementation or evaluation of a control depends on a behavioural change.

5.2 Limitations

Some limitations should be considered when interpreting the present results. This kind of questionnaire does not allow for confident causal conclusions. There were no means to externally validate the responses of the questionnaire. The present results, therefore, may be subject to potential inaccuracies related to the inability of respondents to recall information correctly. First, the main limitation of the current study is its cross-sectional nature. Second, this study exclusively relies on self-report measures. Third, the cross-sectional nature of the survey did not allow for the determination of causal relationships. Fourth, the generalizability of the findings should be taken with caution, as the study was conducted within the dimensions of the Finnish chemical industry. Fifth, longitudinal studies will be needed to establish the causality of the relationships. One cannot be certain that the nature of the responses to the questionnaire would be the same in different countries. The potential lack of measurement calibration would likely have an impact on the relationships between measures as reflected in correlations and regressions.

Longitudinal designs and experimental intervention of OSH studies would make it possible to draw more convincing causal inferences. These kinds of studies could include both the intervention examinations when OSH management and OSH collaboration outcomes would be measured in relation to other OSH variables.

One would predict that a measurement of the OSH behaviour excellence could provide predictive validity about the positive relationship between OSH management and the effectiveness of OSH collaboration.

6 Conclusions and Recommendations

This study indicated that there should be an expanded role for measures of leadership, collaboration, legislation and OHC, e.g. in order to implement the goals of legislation. The following points were emphasized from the results: (1) The two aggregated variables of 'Leadership and Collaboration' and 'Quality of Legislation' had a positive effect on the aggregated variable 'Continuous Improvements', and (2) The two aggregated variables of 'Occupational Health Care' and 'Leadership and Collaboration' exerted a positive effect on the aggregated variable 'Monitoring of Work Environment'. In the climate analysis for the combined variables of 'Monitoring of Work Environment' and 'Risk Prevention Priorities', the following factors were important: Factor 1—Technology and Measurements of Exposure, Factor 2—Management and Guidance, Factor 3—Monitoring and Factor 4—Risk Management.

The present results revealed that good safety performance depends on the commitment of the top management and the collaboration with employees. In dynamic processes such as used by the chemical industry, legislation based on a command-and-control approach is inadequate; instead a fundamentally different view of work system procedure (e.g. self-regulation) is required. The organization needs an informed culture in which those who manage and operate the system have up-to-date knowledge about the human, technical, organizational and environmental factors that determine the organization's OSH as a whole. The present results emphasize the importance of the OHC assessments. For example, the cooperation between the OSH managers and OHC should recommend more frequently the benefits of proactive measures: (1) investigation of OSH, (2) OSH risk assessment, (3) monitoring of OSH, (4) issuing proposals to improve OSH and (5) provision of information, advice and guidance in OSH matters. Safety culture can be improved when the management has a good awareness of OSH. Rather than striving to control OSH behaviour by simply complying with OSH legislation, the focus should be on proactive activities and by encouraging collaboration in OSH. In addition, a safety culture is created as a product of the systemic structure built on the foundations of the joint actions of different parties in the management, collaboration and OHC systems.

Acknowledgments The author is grateful for help with the questionnaire to Kyösti Louhelainen and Maria L. Hirvonen from the Finnish Institute of Occupational Health.

Appendix

Questionnaire to OHS Managers (N = 85) and Workers' OSH Representatives (N = 120) in the Chemical Industry

The following statements were provided and the respondents answered by rating with a Likert scale: (1) Strongly disagree, (2) Somewhat disagree, (3) Somewhat agree, (4) Strongly agree

Q1—Question 1. 'Continuous Improvements'

The following statements were included in this section:

(a) Actual deviations from safety are investigated, for example near accidents, accidents, etc.
(b) The safety aspects have been taken into account in sufficient detail in the instructions for work in order to make sure that work is safe.
(c) The safety practices at my workplace are also applied to subcontractors working in the enterprise.
(d) Pregnant employees are taken into account in the safety practices.
(e) The requirements of the OSH legislation represent the best way to tangibly promote the Continuous Improvements of safety in practice.
(f) The cooperation (between the employer and employees) on OSH matters represents the best way to tangibly promote the Continuous Improvements of safety in practice.
(g) The activities of the management of the enterprise represent the best way to tangibly promote the Continuous Improvements of safety in practice.

Q2—Question 2. 'Leadership and Collaboration'

The following statements were included in this section:

(a) The OSH manager actively promotes the OSH activities at our workplace.
(b) HES (health, environment and safety) management actively promotes the OSH activities at our workplace.
(c) The OSH representative receives appropriate OSH training.
(d) OSH training has been utilized in the OSH activities at my workplace.
(e) The OSH representative receives all necessary information on OSH matters (accidents occurred, the workplace survey drawn up by the OHC service provider, etc.).
(f) The personnel is informed about matters relating to OSH.
(g) We solve and take care of OSH issues together.
(h) Only a few persons take care of OSH matters at my workplace.
(i) The OHC service provider is active in OSH matters at my workplace.

Q3—Question 3. 'Quality of Legislation'
The following statements were included in this section:

(a) The OSH regulations are clear.
(b) The OSH regulations are easy to understand.
(c) The OSH regulations are easy to follow.
(d) The OSH regulations cover very well the various fields of using chemicals.
(e) The OSH regulations increase the employer's interest in the safety and health of personnel.
(f) The OSH regulations are useful for improving employees' own motivation.
(g) The OSH regulations are good but they should be more detailed.
(h) The economic benefits from the OSH regulations are larger than the costs they cause.

Q4—Question 4. 'Monitoring of Work Environment'
The following statements were included in this section:

(a) Work air impurities are followed up through regular measurements of occupational exposure limit (OEL) values.
(b) Any exposure of employees to chemicals is followed up with the help of biological monitoring carried out by the OHC service provider.
(c) It is ensured that occupational exposure level limit (OEL) values known to be harmful are not exceeded.
(d) The training has been arranged to secure that plans and guidance work in practice.
(e) The requirements of the OSH legislation represent the best way to tangibly promote the monitoring of the work environment.
(f) The cooperation (between the employer and employees) on OSH matters represents the best way to tangibly promote the monitoring of the work environment in practice.
(g) The activities of the management of the enterprise represent the best way to tangibly promote the monitoring of the work environment.

Q5—Question 5. 'Occupational Health Care'
The following statements were included in this section:

(a) The OHC service provider has carried out workplace surveys which are utilized when improving the safety.
(b) The OHC service provider is being informed of safety-related changes in the workplace.
(c) Health examinations of employees in the beginning of their employment relationship are carried out, as well as periodical health examinations.
(d) The OHC service provider can be requested for information on health effects of chemicals, when necessary.
(e) Health examinations are organized for employees carrying out work that may cause particular risk of illness.

(f) The requirements of the OSH legislation represent the best way to tangibly promote the OHC services.
(g) The cooperation (between the employer and employees) on OSH matters represents the best way to tangibly promote the OHC services.
(h) The activities of the management of the enterprise represent the best way to tangibly promote the OHC services.

Q6—Question 6. 'Risk Prevention Priorities'
The following statements were included in this section:

(a) The employer has chosen such chemicals for use that cause least the OSH harm.
(b) The employer has chosen such production methods that cause least OSH harm.
(c) My workplace is actively searching for safer alternatives to replace dangerous chemicals.
(d) There is a general air conditioning system that functions well, and the amount of both exhausted air and replacement air are large enough.
(e) Local exhaust ventilation is used in work stations when necessary.
(f) The effectiveness of ventilation is followed up and the ventilation equipment is serviced regularly.
(g) The requirements of the OSH legislation represent the best way to tangibly promote the safe handling of chemicals.
(h) The cooperation (between the employer and employees) on OSH matters represents the best way to tangibly promote the safe handling of chemicals.
(i) The management of the enterprise represents the best way to tangibly promote the safe handling of chemicals.

Q7—Question 7. 'On the On-the-job Training'
The following statements were included in this section:

(a) The line management receives safety training.
(b) The workers receive safety training.
(c) Safety issues form a significant part of employee on-the-job training.
(d) The risks with chemicals and the safe ways to work have been explained to the employees.
(e) The requirements of the OSH legislation represent the best way to tangibly promote the safety training in practice.
(f) The cooperation (between the employer and employees) on OSH matters represents the best way to tangibly promote the safety training in practice.
(g) The activities of the management of the enterprise represent the best way to tangibly promote the safety training in practice.

Q8—Question 8. 'Personal Protective Equipments'
The following statements were included in this section:

(a) The need for personal protective equipment has been defined on the basis of a risk assessment.

(b) The workers participate in the selection of personal protective equipment.
(c) The use of protective equipment is supervised at my workplace.
(d) The storage, maintenance and replacement of protective equipment are duly carried out.
(e) The requirements of the OSH legislation represent the best way to tangibly promote the use of personal protective equipment.
(f) The cooperation (between the employer and employees) on OSH matters represents the best way to tangibly promote the use of personal protective equipment.
(g) The activities of the management of the enterprise represent the best way to tangibly promote the use of personal protective equipment.

Q9—Question 9. 'OSH enforcement operations' (*When you have participated in an OSH inspection*)

The following statements were included in this section:

(a) The OSH enforcement should be restricted only to ensure that the minimum level laid down by law is obeyed.
(b) The OSH inspector should give more advice to workplaces in order to help them to exceed the minimum level laid down by law.
(c) The OSH inspector should more often impose binding obligations to remedy deficiencies in workplaces.
(d) OSH inspectors' visits should be made more often to our workplace.
(e) The OSH inspector should strengthen the enforcement of certain sections of relevant OSH legislation.
(f) The OSH inspector should give more advice.
(g) The OSH inspector should follow up that OSH management systems are being implemented in practice.
(h) The OSH inspector should visit my workplace more often.
(i) The OSH inspector should improve the quality of his/her inspections.

Q10—Question 10. 'Effects of the current OSH inspection' (*When you have participated in an OSH inspection*)

The following statements were included in this section:

(a) During the inspection, the kinds of deficiencies or faults were assessed that would not have been tackled without the inspection.
(b) The inspection triggered the preparation of some OSH documents, or a rewriting of some of some existing documents in a more detailed manner.
(c) The inspection triggered the correction of the deficiencies that had been observed.
(d) During the inspection, we received new information on the legal obligations concerning our workplace.
(e) The inspection led to a more systematic development of OSH.

Q11—Question 11. 'Enforced Policy of the Current OSH Inspection' (*When you have participated in an OSH inspection*)

The following statements were included in this section:

(a) During the inspection, the inspector obtained a truthful picture of our workplace.
(b) All severe hazards and risks at our workplace were not evaluated during the inspection.
(c) The inspection was carried out in a professional manner.
(d) Too much time was used for the inspection.
(e) During the inspection, the employees working at their workstations could express their opinions freely.

References

Ayman R, Adams S (2012) Contingencies, context, situation, and leadership. In: Day DV, Antonakis J (eds) The nature of leadership, 2nd edn. Sage Publications Inc., Thousand Oaks, pp 218–255
Bass BM (1997) Does the transactional/transformational leadership paradigm transcend organizational and national boundaries? Am Psychol 52(2):130–139
Bass BM, Avolio BJ, Jung DI, Berson Y (2003) Predicting unit performance by assessing transformational and transactional leadership. J Appl Psychol 88(2):207–218
Beus JM, Bergman ME, Payne SC (2010) The influence of organizational tenure on safety climate strength: A first look. Accid Anal Prev 42(5):1431–1437
Carmeli A, Friedman Y, Tishler A (2013) Cultivating a resilient top management team: the importance of relational connections and strategic decision comprehensiveness. Saf Sci 51(1):148–159
Cefic (2014) Responsible care. European Chemical Industry Council (Cefic). http://www.cefic.org/en/Responsible-care.html. Accessed 4 June 2014
Colley SK, Lincolne J, Neal A (2013) An examination of the relationship amongst profiles of perceived organizational values, safety climate and safety outcomes. Saf Sci 51(1):69–76
Coyle-Shapiro JAM, Shore LM (2007) The employee-organization relationship: where do we go from here? Hum Resour Manag Rev 17(2):166–179
Cunliffe A (2008) Orientations to social constructionism: relationally responsive social constructionism and its implications for knowledge and learning. Manag Learn 39(2):123–139
Dansereau F, Graen G, Haga WJ (1975) A vertical dyad linkage approach to leadership within formal organizations: a longitudinal investigation of the role making process. Organ Behav Hum Perform 13(1):46–78
Díaz-Cabrera D, Hernández-Fernauda E, Isla-Díaz R (2007) An evaluation of a new instrument to measure organisational safety culture values and practices. Accid Anal Prev 39(6):1202–1211
Dienesch RM, Liden RC (1986) Leader-member exchange model of leadership: a critique and further development. Acad Manag Rev 11(3):618–634
Fairhurst G, Grant D (2010) The social construction of leadership: a sailing guide. Manag Commun Quart 24(2):171–210
Finnish Legislation (2001) Finnish Occupational Health Care Act (1383/2001). Helsinki. http://www.finlex.fi/fi/laki/kaannokset/2001/en20011383. Accessed 16 June 2014
Finnish Legislation (2002) Finnish Occupational Safety Act (738/2002). Helsinki. http://www.finlex.fi/fi/laki/kaannokset/2002/en20020738.pdf. Accessed 16 June 2014

Finnish Legislation (2006) Finnish act on occupational safety and health enforcement and cooperation on occupational safety and health at workplaces (44/2006). Helsinki. http://www.finlex.fi/fi/laki/kaannokset/2006/en20060044.pdf. Accessed 16 June 2014

Grote G (2012) Safety management in different high-risk domains—all the same? Saf Sci 50 (10):1983–1992

Grote G, Künzler C (2000) Diagnosis of safety culture in safety management audits. Saf Sci 34 (1–3):131–150

Hale A, Borys D (2014a) Working to rule, or working safely? Part 1: a state of the art review. Saf Sci 55(1):207–221

Hale A, Borys D (2014b) Working to rule or working safely? Part 2: the management of safety rules and procedures. Saf Sci 55(1):222–231

Hale AR, Heming BHJ, Catfhey J, Kirwan B (1997) Modelling of safety management systems. Saf Sci 26(1/2):121–140

Hersey P, Blanchard KH (1982) Management of organizational behavior: utilising human resources. Prentice Hall International, London

Huy QN (2002) Emotional balancing of organizational continuity and radical change: the contribution of middle managers. Adm Sci Q 47(1):31–69

ICCA (2014) Responsible care. International Council of Chemical Associations (ICCA). http://www.icca-chem.org/. Accessed 16 June 2014

Kaiser HF (1974) An index of factor simplicity. Psychometrika 39(1):31–36

Kapp EA (2012) The influence of supervisor leadership practices and perceived group safety climate on employee safety performance. Saf Sci 50(4):1119–1124

Le Coze J-C (2013) Outlines of a sensitising model for industrial safety assessment. Saf Sci 51 (1):187–201

Likert R (1961) New patterns of management. McGraw-Hill, New York

Locke EA, Latham GP (1990) A theory of goal setting and task performance. Prentice-Hall, Englewood Cliffs

Locke EA, Latham GP (2002) Building a practically useful theory of goal setting and task motivation: a 35-year odyssey. Am Psychol 57(9):705–717

Lord RG, Din JE (2012) Aggregation processes and levels of analysis as organizing structures for leadership theory. In: Day DV, Antonakis J (eds) The nature of leadership, 2nd edn. Sage Publications Inc., Thousand Oaks, pp 29–65

Luria G, Yagil D (2010) Safety perception referents of permanent and temporary employees: Safety climate boundaries in the industrial workplace. Accid Anal Prev 42(5):1423–1430

Morrow SL, McGonagle AK, Dove-Steinkamp ML, Walker JrCT, Marmet M, Barnes-Farrell JL (2010) Relationships between psychological safety climate facets and safety behavior in the rail industry: a dominance analysis. Accid Anal Prev 42(5):1460–1467

Olsen E (2010) Exploring the possibility of a common structural model measuring associations between safety climate factors and safety behaviour in health care and the petroleum sectors. Accid Anal Prev 42(5):1507–1516

OSHA (2012) Workers' Rights: OSH Act 1970. U.S. Department of Labor. Occupational Safety and Health Administration. https://www.osha.gov/Publications/osha3021.pdf. Accessed 16 June 2014

Piccolo RF, Colquitt JA (2006) Transformational leadership and job behaviors: the mediating role of core job characteristics. Acad Manag J 49(2):327–340

Reason J (1997) Managing the risks of organizational accidents. Ashgate Publishing Company, Aldershot

Reniers G (2009) An optimizing hazard/risk analysis review planning (HARP) framework for complex chemical plants. J Loss Prev Process Ind 22(2):133–139

SAS (2005) SAS user's guide. Version 9. SAS Institute Inc., Cary

Schriesheim CA, Castro SL, Cogliser CC (1999) Leader-member exchange (LMX) research: a comprehensive review of theory, measurement, and data-analytic practices. Leadersh Quart 10 (1):63–113

Schriesheim CA, Castro SL, Yammarino FJ (2000) Investigating contingencies: an examination of the impact of span of supervision and upward controllingness on leader-member exchange using traditional and multivariate within- and between-entities analysis. J Appl Psychol 85 (5):659–677

Uhl-Bien M, Maslyn J, Ospina S (2012) The nature of relational leadership: a multitheoretical lens on leadership relationships and processes. In: Day DV, Antonakis J (eds) The nature of leadership, 2nd edn. Sage Publications Inc., Thousand Oaks, pp 289–330

Vroom VH, Yetton PW (1973) Leadership and decision-making. University of Pittsburgh Press, Pittsburgh

Wassenaar CL, Pearce CL (2012) The nature of shared leadership. In: Day DV, Antonakis J (eds) The nature of leadership, 2nd edn. Sage Publications Inc., Thousand Oaks, pp 363–389

Wirth O, Sigurdsson SO (2008) When workplace safety depends on behavior change: topics for behavioral safety research. J Saf Res 39(6):589–598

Yukl GA (1999) An evaluation of conceptual weaknesses in transformation and charismatic leadership theories. Leadersh Quart 10(2):285–305

Yukl GA, Becker WS (2006) Effective empowerment in organizations. Organ Manag J 3(3):210–231

Zohar D, Tenne-Gazit O (2008) Transformational leadership and group interaction as climate antecedents: a social network analysis. J Appl Psychol 93(4):744–757

Chapter 13
Discourses of the Different Stakeholders About Corporate Social Responsibility (CSR)

Toivo Niskanen

Abstract The purpose of this study was to examine the concept of corporate social responsibility (CSR) as viewed by different participants in the Finnish labour market. One senior specialist from each of the Confederation of Finnish Industries, the Central Organization of Finnish Trade Unions SAK and the Finnish Consumers' Association was requested to complete an online questionnaire. The aim of this study was to elucidate how the different participants view various aspects of CSR. Further study aims were to gather the kinds of the responses to the research questions which could be used as a reference base for future development. The main research question was: How does CSR become the best practice in business activities? The sub-questions were: How does CSR come to represent the best practices in the day-to-day activities of (1) business strategy of companies, (2) financiers and shareholders, (3) customers and consumers, (4) employees of the companies and (5) communities and authorities? CSR is a strategic process of mutual commitment between a corporation and its stakeholders. Its purpose is to create a social contract to maximize the welfare of all partners concerned. CSR is also an important way to stimulate proactive behaviour and to improve to health, environment and safety strategies of the companies. CSR provides a company with the ability to recognize and respond effectively to social challenges.

Keywords Corporate social responsibility · CSR · Sustainability · Environment · Safety · Discourse analysis

T. Niskanen (✉)
Ministry of Social Affairs and Health, Occupational Safety and Health Department, Legal Unit, P.O. Box 33, FI-00023 Helsinki, Finland
e-mail: toivo.niskanen@stm.fi

1 Introduction

This study of current corporate social responsibility (CSR) consists of perceived responsibilities of CSR and how these are associated with various types of CSR practices, the drivers and motivators that influence decisions to engage in such practices, and the underlying relationship between business and society that is being increasingly emphasized. Defining CSR is challenging, as there are many, sometimes conflicting definitions that attempt to explain its governing concept(s) using normative social constructions that can vary across cultures, etc. Most definitions of CSR describe it as constituting actions whereby companies integrate societal concerns into their business policies and operations. Wood (1991) presents an integrated definition of CSR as the configuration of the principles of social responsibility, processes of social responsiveness, and policies, programmes, and observable outcomes as they relate to the firm's societal relationships. Briefly, a firm committed to CSR has principles and processes in order to minimize its negative impacts and maximize its positive impacts on societal at all or on selected groups (Wood 1991). The European Union (European Parliament Resolution 2006) defines CSR as a concept whereby companies integrate social and environmental concerns in their business operations and in their interaction with stakeholders on a voluntary basis. CSR from the managerial perspective can lead to a better balancing of corporate objectives and societal risks; from the regulatory perspective, it offers the prospect of reflexive types of regulation; and from the financial perspective, it holds out the possibility of new types of deliberation, based around shareholder engagement with companies (Deakin and Hobbs 2007).

There are many viewpoints, however, as to whom or what constitutes a stakeholder and whether or not they should even be considered in the decisions of the corporation. Furthermore, there are also many potential stakeholders, including individuals and groups in the workplace (employees), the marketplace (customers, suppliers), government, and the community; as well as non-governmental organizations (NGOs) with very specific interests such as the environment, ethics and human rights (Moir 2001). The idea of addressing any perceived responsibilities to stakeholder(s) other than shareholders by engaging in CSR-type activities is viewed by some as a violation of free-market principles. For instance, Milton Friedman (1970) wrote an opinion piece in the New York Times that focused on debunking the validity of companies engaging in anything other than strategies that will increase their profits. According to Friedman, businesses and markets are amoral, and as a result should not venture into what he considered social policy. He surmises that issues of morality should be dealt with in the political realm. As far as the market and businesses are concerned, as long as everyone is operating in their own best interest, free of coercion, and with access to full information, the benefits will balance. A number of responsibilities are identified in the literature include the following examples by Moir (2001): (1) Treat employees fairly and equitably; (2) Operate ethically and with integrity; (3) Respect basic human rights; (4) Sustain the environment for future generations; and (5) Be a caring neighbour in their communities.

Finland's national strategy for sustainable development was adopted in June 2006 by the Finnish National Commission on Sustainable Development. One of the ultimate aims is to consider the impacts of sustainable development in an integrative way and, consequently, to encourage the creation of the potential win–win–win opportunities for developing a sustainable society. The approach is already referred to as the 'Finnish model', in which broad-based, multi-stakeholder participation is combined with ministerial-level political leadership. According to the Finnish strategy, success in a changing world requires that Finland will develop further as a knowledge and innovation-based society which promotes the utilization and development of national strengths: education and know-how, technology, good governance, equality and a high level of environmental protection. This strategy and the policies aimed at sustainable development particularly strengthen the innovation process and empower citizens to take a strong role in developing society (Finnish National Commission on Sustainable Development 2006).

The Finnish innovation system is a broad-based collective that is formed by producers and utilizers of new knowledge and expertise as well as the interactive relationships between these two participants. A wide range of policies—originating from EU, national and local levels—have to be fine-tuned to support innovation. Business activities, education, R&D operations, a knowledgeable workforce and funding are all key components of this innovation system (Confederation of Finnish Industries EK 2014b). Universal social protection and extensive welfare services represent the foundation stones constitute of the Finnish welfare society. The Confederation of Finnish Industries EK is the leading business organization in Finland. It represents the entire private sector, both manufacturing industry and service sectors, and companies of all sizes. EK's member companies account for more than 70 % of Finland's gross domestic product and over 95 % of Finnish exports. There are many challenges which can be grouped under the umbrella of CSR e.g. as follows: (1) global competitiveness, (2) developing responsible practices throughout the whole supply chain, (3) well-being and skill base of personnel of the companies, (4) health and safety, (5) combating climate change, (6) eco-efficient use of raw materials and (7) cooperation with main stakeholders (Confederation of Finnish Industries EK 2014c).

The Finnish labour market system is characterized by prominence of labour market organizations not only within the labour market itself, but wider throughout society, since there has been extensive tripartite cooperation between the national government and these social partners (employers and trade unions). The regulation of the labour market in Finland is based on labour legislation primarily on collective agreements. Any trade union and employers' association may make collective agreement. The Finnish main employers' organization has stated that the challenges to the national economy in the long term relate to the ageing population and the availability of skilled labour (Confederation of Finnish Industries EK 2014a). Both employees and employers in Finland are highly organized. This high level of organization makes it possible for national employer and labour confederations to negotiate very broad collective agreements, which then serve as guidelines for collective bargaining by individual industrial trade unions and employers' federations.

The national government also consults the trade unions and the employers over any proposed amendments to the laws that affect working life. The rights of employees working in Finland are guaranteed by law and by collective agreements negotiated by the trade unions on behalf of employee groups (Finnish Trade Union Movement 2014).

For example, the goal of the strategy of the Ministry of Social Affairs and Health (Finnish Strategy 2011, p. 3) is to achieve a socially sustainable society in which individuals are treated equally, everyone has the opportunity to participate, and everyone's health and functional capacity is supported. Since monitoring may be considered a form of direction, its predictive role must be enhanced (Finnish Strategy 2011, p. 16). There are three strategic choices concerned with OSH: (1) A firm foundation for welfare; (2) Access to welfare for all; and (3) A healthy and safe living environment. In the Finnish Strategy (2011, pp. 6–7), the shift of the focus is moving from treating the sick to actively promoting well-being with the overall aim being to lengthen working careers by an average of 3 years by 2020. The attractiveness of working life must be increased by improving working conditions and well-being at work. The improvement in working conditions must be achieved through joint efforts by management and personnel. The risk of permanent working incapacity must be reduced by promoting health and work ability, by improving working conditions and by enhancing OHC (Finnish Strategy 2011, p. 8).

2 The Aim of the Study and the Research Questions

The aim of this research was to study what kind of view different industrial stakeholders have on CSR. A further aim of the study was to collect material to serve as a reference for a future project being developed to examine this topic.

The main research question is: How does the CSR become the best practice in business activities?

The sub-questions for this research are as follows: How does the CSR become the best practices in the day-to-day activities of (1) business strategy of companies, (2) financiers and shareholders, (3) customers and consumers, (4) employees of the companies and (5) communities and authorities?

3 Literature Review and Practical and Theoretical Framework

3.1 CSR Dimensions in the Tripartite Instruments

When developing their own CSR approaches, businesses are guided by standards and principles derived from ILO, UN and OECD conventions and other legislation which has been adopted at some multilateral level through an intergovernmental process after consultation with business, labour and other stakeholders (OECD 2005; ILO 2006; United Nations 2007). Codes of conduct are directive statements which provide guidance and prohibit certain kinds of conduct. Some are intended to guide a company's own environmental and social impacts; others focus on the impacts of their suppliers; others may apply to both groups (OECD-ILO 2008, p. 6). Enterprises have to operate in the framework of legislation, regulations and administrative practices in their home countries, and in many cases they need to consider of relevant international agreements, principles, objectives, and standards, to ensure that their activities protect the environment, public health and safety (OECD 2005, p. 41).

The Commission of the European Communities (2002, 2006) has provided some guidelines about CSR. In its Communication published in 2002, the Commission proposes to build its strategy to promote CSR on a number of principles, e.g. ensuring compliance and compatibility with existing international agreements and instruments (e.g. ILO, OECD and UN) for multinational enterprise. In a subsequent Communication (2006), the Commission stated that it is committed to promoting awareness and implementation of these instruments and furthermore, it will work together with other governments and stakeholders to enhance their effectiveness. The European Parliament (EP) Resolution (2006) on CSR makes clear its support for the EU's Eco-Management and Audit Scheme and stresses its scope for developing similar schemes concerning the protection of labour, social and human rights, Furthermore, the EP emphasized that the CSR debate must not be separated from questions of corporate accountability, for example enterprises must consider issues of the social and environmental impact of their business activities, as well as their relations with stakeholders, the protection of minority shareholders' rights and the duties of company directors.

ISO Standard 26000:2010 'Guidance on Social Responsibility' is being focused to help an organization achieve mutual trust with its stakeholders by improving its social responsibility performance. An organization's social responsibility performance may influence: (1) the general reputation of the organization; (2) its ability to recruit and retain a committed workforce and/or members, but also to attract customers, clients or users; (3) the maintenance of their employees' morale and productivity; (4) the views of investors, donors, sponsors and the financial community; and (5) its relationship with government, the media, suppliers, peers, customers and the community in which it operates. To assess its social responsibility, an organization needs to consider the following core issues: (1) organizational governance; (2) human rights; (3) labour practices; (4) the environment; (5) fair operating practices; (6) consumer issues; and (7) contribution to the community and society (ISO 26000:2010).

3.2 The Framework of the Study

In a world of complexity and change, managers are asked to tackle a much greater diversity of CSR problems. They have to continue to ensure that organizational CSR processes are efficient and that they are served by the latest developments in technology (applied from Jackson 2003). Jackson (2003) stated that with increasing complexity, change and diversity, managers have inevitably sought the receive help of the following: scenario planning; benchmarking; value chain analysis; continuous improvement; total quality management; learning organizations; process re-engineering; and knowledge management.

Senge (1994) regards system dynamics, presented as 'the fifth discipline', as the most important tool that organizations must master on the route to becoming 'learning organizations'. Only system dynamics can reveal the systemic structures that govern their behaviour. Nevertheless, it is essential to support study of the fifth discipline with research on the other four disciplines seen as significant in the creation of learning organizations. These are 'personal mastery', 'managing mental models', 'building shared vision' and 'team learning'.

French (1993, pp. 228–235) concludes that a company has an identity equivalent, in a moral sense, it can be considered as the equivalent of an individual being and thus it has its own ethical responsibility. Similarly, Freeman and Liedtka (1997, pp. 286–296) emphasized that a company must assume ethical responsibilities for all those individuals and groups with which the company operations impact. Carroll (2000a, b) defines the model of the company's responsibilities to be the implementation of philanthropy, ethical, legal and economic responsibilities. In this model, philanthropy is the desired objective, ethics is the expected target, and legality and economic goals are the required business objectives. The other scholars (e.g. Velasques 1992, p. 16; Cavanagh et al. 1995, p. 198) on business ethics have come to similar conclusions when they have reviewed a company's interaction with its surrounding society.

Freeman and Liedtka (1997, pp. 286–296) state that a company's social role is to use the available resources as efficiently as possible. The principles can also be applied to the different dimensions of the business, e.g. competitiveness, leadership, the responsibility of the multinational company and the business responsibilities in the society. In this respect, CSR is an element which impacts on company's financial results and has to be targeted to the entire chain of supply.

Elkington (2001) considered that the multidimensional parameters were important in CSR. In this sense, several dimensions can be considered to have the widest impact, e.g. economic well-being, environmental management and social justice. Furthermore, when social responsibility has been integrated into the everyday business routines it achieves beneficial results, even if the underlying principles are not necessarily clearly appreciated by the organization's members (e.g. Catasus et al. 1997; Crane 2000). Elkington (2001) also identified the importance of the economic aspect linked to human capital and knowledge capital. However, the environment management pursued by companies has also attracted criticism for being superficial with a tendency to adopt only cosmetic improvements (e.g. Welford 2000, p. 133; Crane 1999) even though environmental values tend to be accorded high importance in annual reports. Furthermore, moral choices have been found to be rated as of rather minor importance even among the most environment sensitive companies (e.g. Bansal and Roth 2000; Crane 2000; Fineman 1996, 1997), and in turn the need to comply with legislation is often found to be the strongest and most effective motivator in sustainable development (Bansal and Roth 2000).

The term 'stakeholder' can be defined broadly by a holistic and interdependent view defining stakeholders as 'any group or individual who can affect or are affected by the achievement of the organization's objectives' (Salzmann et al. 2006, p. 3). Thus, the stakeholder concept includes media, trade unions, public authorities, non-governmental organizations, communities, etc. (Freeman and Liedtka 1997). When considering sustainable development and CSR, then the stakeholders' contributions to companies involve both tangible and intangible resources: shareholders provide capital; employees offer labour; customers have loyalty and a certain willingness to pay; and communities and governments are expected to establish a stable and beneficial regulatory framework, etc. (Salzmann et al. 2006, p. 3) In turn, the inducements for companies are similarly diverse: they can range from a certain return on investment to safe working conditions and high-quality products (Salzmann et al. 2006, p. 3).

Dobers and Wolff (2000) have argued that the holistic business approach will create sustained success for a company if it is successful in balancing stakeholders' demands and sustainability requirements. Nonetheless, this balancing process is not easy; the management of a business has to be able to deal with value concepts that are conflicting and inherently contain trade-offs (Dobers and Wolff 2000). Furthermore, some stakeholders can have an indirect effect on corporations via other stakeholders. Ramus and Montiel (2005) argue that the institutional environment (e.g. pressures from stakeholders or government) creates incentives for companies to commit to very similar types of environmental policies regardless of the type of industry in which they do business. Salzmann et al. (2006) defined corporate sustainability performance as a configuration of three components: (1) four determinant drivers (issues, stakeholders, managers and company-specific drivers); (2) corporate sustainability management (as a profit-driven corporate response to social and environmental issues); and (3) the resulting outcome, characterized by social and environmental impacts and associated changes in financial performance.

Adapted from ethics described by from Fisher and Lovell (2006, p. 209) to CSR the pressures that work with the many issues tend to do so from one of three perspectives, as reflected in the column on the right-hand side of Fig. 1. These levels are the macro and global and societal level, the corporate and organizational level and the micro and individual level (applied from Fisher and Lovell 2006, p. 209).

3.2.1 Business Strategy of the Companies

CSR stresses the importance of clearly acknowledging economic responsibility, i.e. the long-term financial health of the company represents the fundamental organizing principle of companies (e.g. Salzmann et al. 2005). However, if sustainability is not an explicit part of the strategic business plans of a company, then there is empirical evidence revealing that companies are unlikely to actually implement

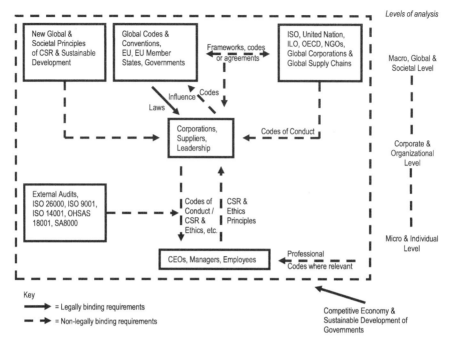

Fig. 1 Formal and informal pressures for CSR (adapted from Fisher and Lovell 2006, p. 207)

sustainability programmes (Ramus 2005; Ramus and Montiel 2005; Ramus and Killmer 2007). The top management defines the appropriate strategic response to a given event in order to save costs, to minimize risk to the company is reputation, to comply with the laws of the country but if possible to gain a competitive advantage (Ramus and Oppegaard 2006, p. 12). For example, not only an organization's specific field of industry, but also the culture, values, history, management philosophy, top management commitment, control systems and monitoring mechanisms may affect its approach to managing sustainability issues (Ramus and Oppegaard 2007, p. 5).

3.2.2 Financiers and Shareholders

Financial institutions are crucial facilitators for sustainability. Financial actors demand business' transparency in management systems and it has been predicted that objectives for sustainability will play an important role in the near future (Dobers and Wolff 2000). Steger (2006, p. 4) concluded that nowadays the financial institutions exert a tremendous influence on enterprises and that this applies in particular to the financial mainstream, which is not a strong driver for corporate sustainability. Furthermore, although socially responsible investing and corporate sustainability ratings and indices are becoming publicized, they still have a rather insignificant overall impact (Steger 2006, p. 4). Financial institutions also assess industries from an environmental point of view; they judge and evaluate environmental risks and thus price companies realistically depending on their liabilities, because the environmental risks taken by companies can have a negative impact on their longer-term shareholder value (Dobers and Wolff 2000). Similarly, Salzmann et al. (2005) concluded that corporate sustainability management manifests itself through corporate social and environmental initiatives undertaken to exploit financial opportunities and to minimize financial risks by mitigating or resolving those issues. Pirson and Malhotra (2008, pp. 11–12) stated that internal stakeholders, such as investors, are most often seeking for evidence of managerial competence: they want to have confidence in the ability of management to effectively control costs and to manage the workforce effectively so that they remain competitive and create financial value.

Orlitzky et al. (2003) conclude that market forces generally do not penalize companies that are high in corporate social performance and thus, managers can afford to be socially responsible. Their results of the meta-analysis revealed a positive association between social/environmental performance (CSP) and corporate financial performance (CFP) across industries and across study contexts. If managers believe that CSP is a prerequisite for CFP, they may actively pursue CSP in the belief that the market will reward them for such efforts. The company's executives and top-leadership must be attentive to the perceptions of third parties, regardless of whether they are market analysts, public interest groups or the media (Orlitzky et al. 2003).

Salzmann et al. (2006, p. 15) have identified a broad range of possible stakeholder impacts, e.g.: (1) impacts resulting in the innovation of processes, products and supply chains and (2) impacts on brand value and reputation and customer loyalty. Similarly, it has been claimed that shareholder value will be best realized if companies include environmental and social responsibility objectives in their value creating statements, processes and products (Dobers and Wolff 2000). Stakeholders of all types are interested in associating with organizations with whom they can identify, and with whom they perceive a match in values (Pirson and Malhotra 2008, p. 17).

In general, adopting a proactive corporate environmental strategy that goes beyond regulatory compliance can have a positive effect on corporate financial performance (e.g. Christmann 2000; Hart 1995; Wagner 2005). The size of the enterprise has a significant effect on the degree of proactiveness, with larger organizations being more likely to adopt proactive environmental practices (Sharma 2000). However, Steger (2006, p. 5) has stated that the economic logic and business advantages of corporate sustainability are not strong enough and that although and assessment of corporate social performance may occur, it is still not a part of the mainstream analyses. Dobers and Wolff (2000) detected a strong trend in that recent strategic assessments have started to include sustainability criteria in addition to the more conventional financial aspects.

3.2.3 Customers and Consumers

The external stakeholders, e.g. customers and suppliers, typically careless about managerial competence but much more about technical competence: the ability to produce goods and services of high quality and to deal effectively with the supply chain (Pirson and Malhotra 2008, pp. 11–12). Furthermore, Dobers and Wolff (2000) evaluated that the concept of 'environmental awareness' in society. They indicated that consumers behave according to their ethical values, which would imply that they are more likely to purchase environmentally friendly products when these are available.

Furthermore, Dobers and Wolff (2000) concluded that when looking at segmentation and marketing strategies, there is a strong indication that the 'green' consumer differs considerably from other customers. They believed that this means from a business marketing point of view. A company may have to adopt a differentiated marketing strategy (Dobers and Wolff 2000). There is an OECD recommendation that when dealing with consumers, enterprises should act in accordance with fair business, marketing and advertising practices and should take all reasonable steps to ensure the safety and quality of the goods or services they provide (OECD 2005, p. 25). Furthermore, Steger (2006, p. 5) indicated that consumer organizations are in some way 'happy' that companies (particularly those that have branded goods and are close to the consumer) are making a contribution to societal well-being.

Steger (2006, pp. 5–6) claimed that consumer organizations lack resources because their activities are largely based on their own product-related research. While consumers seem to express positive environmental attitudes, when it comes to their behaviour, the challenges faced by companies are much more ambiguous (Dobers and Wolff 2000). Corporate customers also have a clear business case to ensure that their suppliers are complying with certain social and environmental standards (Steger 2006, p. 6). In contrast, corporate suppliers often do not 'dare' to bother their customers with corporate sustainability, unless there is a clear need for product responsibility and risk management, e.g. hazardous chemicals (Steger 2006, p. 6).

3.2.4 Employees of the Companies

Workplace CSR projects can affect many different areas of a company's human resource policy such as health and safety, the work-life balance of employees, staff diversity and cultural awareness. Many enterprises with a demonstrated capability of developing a shared vision seem to be able to accumulate the skills necessary for developing a proactive environmental strategy much earlier than companies without this kind of shared vision because these strategies depend 'upon tacit skill development through employee involvement' (Hart 1995). When the enterprise decides to integrate the environmental issues into the organization, it can exploit certain resources and capabilities developed by the company in its business management that can be viewed as 'potential environmental capabilities' (Clavera et al. 2007; Hart and Ahuja 1996). The corporate environmental strategy places an enterprise's environmental strategies on a spectrum ranging from reactive strategies to more proactive strategies that includes voluntary eco-efficient practices in environmental leadership strategies. In this latter strategy, the products, processes and even business models are redesigned along the entire product life cycle (Sharma 2000; Sharma and Vredenburg 1998).

3.2.5 Communities and Authorities

Ramus and Montiel (2005) have argued that the institutional environment (e.g. pressures from communities and government) represent incentives for firms to commit to very similar types of environmental policies regardless of the type of industry in which they do business. A central consideration for business organizations must be how well private CSR initiatives reflect and reinforce government agreements on labour, social and environmental standards (OECD-ILO 2008, p. 13). Steger (2006, p. 5) indicated that most communities constantly struggle to maintain employment and that there is strong competition between communities throughout Europe. Local communities and governments are primarily concerned about employment levels and thus regional competitiveness. In this situation, higher social and environmental standards are largely 'taboo', and they revert to an enabling approach to corporate sustainability (Steger 2006, p. 5). Enterprises are encouraged to cooperate with local communities and government authorities in the development and implementation of policies. Taking into consideration, the views of other stakeholders in society including the local community and governmental interests can enrich this process (OECD 2005, p. 41). Steger (2006, p. 5) concluded that in light of the weakening bargaining power of the governments (against the major multinationals in particular), they have also tried to target the reputation of companies that have decided to relocate their manufacturing base or lay-off significant numbers of their workforce.

A value-based approach involves open communication and dialogue with NGO associations. The outcomes of this approach can lead to sustainability and active participation by the company in the community (Ramus and Oppegaard 2006, p. 11). Enterprises need to take fully into account established NGO policies in the countries in which they operate, and to consider the views of other stakeholders. In this regard, it is recommended that enterprises should encourage local capacity building through close cooperation with the local community (OECD 2005, p. 25). There are many ways to increase the vitality of rural areas, e.g. by providing support for entrepreneurship, promoting a diverse business structure and utilizing the particular strengths of each region as well as promoting their sense of local community and local culture. Many rural and urban communities are struggling to create sustainable change for themselves—and this challenge is heightened because many of them do not have access to the professional skills needed to help overcome deprivation and social exclusion (Business in the Community 2008). By working together, businesses and communities can create exciting solutions that will be beneficial to both parties (Business in the Community 2008).

4 Research Methods and Research Materials

4.1 Constructionism and Discourses

The data from this discourse study were produced in March and April in 2008. It originated from an online questionnaire form which included open-ended theme questions classified in different taxonomies. The material on CSR was answered by the senior specialists from three organizations, the Confederation of Finnish Industries EK, the Central Organization of Finnish Trade Unions SAK and the Finnish Consumers' Association.

Constructionism starts with the assumption that access to shared dynamic, changing and individually constructed reality only occurs through social constructions such as language and shared meanings in everyday interactions (Lord and Dinh 2012). From this perspective, knowledge is socially constructed: social reality, identities and knowledge are culturally, socially, historically and linguistically influenced. In this method, the social construction approach focuses on relationality (Cunliffe 2008); CSR is being constructed 'in relation to some goal'. The CSR relationships are emergent and co-constructed as an interactive dynamics (applied from Fairhurst and Grant 2010). Here the interests within the processes constitute CSR as the outcome of particular types of relationships and interactions.

Discourse analysis (DA) is a term for a broad area of language study, containing a diversity of approaches with different epistemological roots, and very different methodologies. In general, DA can be defined as a 'set of methods and theories for investigating language in social contexts' (Wetherell et al. 2001; Wetherell 1998). From this perspective, discourse is not 'merely about actions, events and situations, it is also a potent and constitutive part of those actions, events and situations' (Potter 1997, p. 144; Potter 2000, p. 31). DA can be thought of as focusing on the production of versions of reality and cognition as parts of practices in natural settings. DA is concerned with understanding the nature of power and dominance and how discourse contributes to their production (van Dijk 1995, 2001, pp. 421–434; Fairclough 1995, 1996, 2000; Potter 1996). There are four broad sets of research issues which can productively be addressed specifically by discourse analysts in transdisciplinary research on organizational change: the problems of emergence, hegemony, recontextualization and operationalization (Fairclough 2005).

For example, in CSR, these mediating entities are CSR practices, articulations of diverse CSR elements, discourses which constitute CSR selections as concrete entities in particular areas of social strategy. Translation is a concept that captures the process of spread, and explains how micro-level activities contribute to institutional change. Furthermore, even though translations must be understood at a

Fig. 2 The paradigm model (Strauss and Corbin 1998, p. 127)

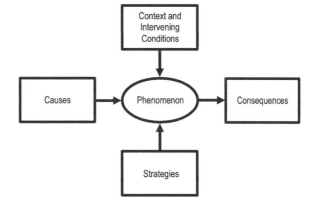

local level, they can occur globally and among groups of organizations, thereby contributing to institutional change (Windell 2006, pp. 28–29). Czarniawska and Sevon (1996) introduced translation as a concept to explain how organizational change emerges as management ideas and practices are translated into actions that may eventually be institutionalized in an organization or in larger groups of organizations. Tsoukas and Chia (2002) recognized that the categories and practices which are institutionalized in organizations are inevitably subject to adaptation and change as organizational agents engage in a range of processes and events.

In order to formulate the relationships, Strauss and Corbin (1998, p. 127) suggested a coding paradigm model, which is symbolized in Fig. 2.

4.2 Classification of Emergence, Hegemony, Recontextualization and Operationalization

Windell (2006, pp. 28–29) concluded that the translation in CSR can be understood as contributing to institutional change if practices, ideas and objects are replicated and stabilized into institutions. There are four broad sets of research issues which can productively be addressed specifically by CSR discourse in DA research: the problems of emergence, hegemony, recontextualization and operationalization (applied from Fairclough 2003):

- *Emergence* is the process of emergence of new CSR discourses and their constitution. New CSR discourses emerge through 'reweaving' relations between existing CSR discourses. These may include 'external' discourses existing at the EU, ILO and UN level which become recontextualized with the social partners' agents.
- *Hegemony* is the process of particular emergent CSR discourses becoming hegemonic in particular social partners' organizations. The effect of emergent

CSR discourses depends upon whether they are selected for incorporation into the strategies of social partners.
- *Recontextualization* is the dissemination of hegemonic CSR discourses across structural boundaries (e.g. between social partners' organizations) and scalar boundaries (e.g. from CSR international scale to national, or to local). It involves reception and appropriation, working 'international' CSR discourses into relations with 'national' discourses.
- *Operationalization* is the enactment of CSR discourses in new ways. It involves (inter)acting of social partners' discursions and their materialization as practices of the business society.

The 'flow' of CSR discourses across scalar boundaries between 'global' (UN, OECD, ILO), 'international' (e.g. EU) and 'national', is conditional upon how they enter into 'national' social partners' relations.

The construction of ideas in CSR can be understood as activities of translation, in which ideas are objectified into new forms and practices that fit the local institutional context (Windell 2006, p. 41). This research is based upon the theoretical claim that CSR discourses are elements of social partners' business life, and they are dialectically interconnected with other elements, and may have constructive and transformative effects on other elements of the business life. It also claims that CSR discourses have in many ways become more potent as elements of business life.

The discourse classification of RC programmes bases on the Fairclough's (1996, p. 64) constructivism theoretical framework. DA's results do not merely describe the world, they also are expressing, synthesizing and building social identities, relations between human beings and information systems (Fairclough 1996, pp. 64–66). Key assumptions can be summed up as follows: (1) discourses of the company reflect a practical reality, (2) business experts can use a range of discourses, which can be parallel and compete with each other, (3) the practical action is context bound and (4) discourse is built on strategies and policies. Discourse is a dual concept. Firstly, it means the interactive process in which implications are produced. Secondly, it refers to the final outcome of this process.

4.3 Classification with the Actantial Model

In the data analysis of the discursion in the online questionnaire, the actantial model of Greimas (1987, 1990) was applied.

The six actants are divided into three oppositions, each of which forms an axis of the actantial description (Hébert 2011):

- The axis of goal: (1) subject/(2) object. The subject is what is directed towards an object. The relationship established between the subject and the object is called a junction.
- The axis of power: (3) helper/(4) opponent. The helper assists in achieving the desired junction between the subject and object; the opponent hinders this goal.

- The axis of transmission and knowledge: (5) sender/(6) receiver. The sender is the element requesting the establishment of the junction between subject and object. Sender elements are often receiver elements as well.

The core of the actantial model is formed from the subject's intention to acquire the object. Consequently, the subject is the performer of a task, and the object is something that the subject tries to acquire through some activity (Greimas 1987). A sender is a party who gives the task to the subject (Greimas 1987, 1990). At the same time, the sender defines the value targets for the activity, and provides justification and motivation for the subject's activity. A receiver is a party who evaluates whether the subject has succeeded, and confers on the parties to the discourses either a reward or punishment after the task has been completed.

5 Results of the Discourses of the Labour Market Organizations and Stakeholders

5.1 Business Strategy of Companies

It is also essential that the wider impacts and risks are considered when strategic business decisions are made. In CSR, the confidence between the companies and the stakeholders increases and also unexpected effects on the business image may be reduced. Ethical investment offers investors something extra over and above conventional collective investment. An ethical investment is an investment which explicitly seeks to take into account environmental, social and ethical issues. It applies the combination of financial and ethical objectives to the selection of investments.

Figure 3 presents the semiotic actantial model developed by Greimas (1987, 1990) which here is used to link the theory and the discourse data of CSR in relation to the business strategy of the companies.

5.2 Financiers and Shareholders

Emergence: Companies are committed to creating long-term value for their shareholders, recognizing that sustainable profit is essential for the continuity of their business. Companies provide competitive returns on their shareholders' investments. In this respect, they have a responsibility to take due account of the expectations of their investors. A growing number of companies are reporting about their CSR activities. As a result, companies have a variety of new needs in terms of information and performance management, e.g. social reporting, management of human capital, etc. The requirements for CSR data have become progressively more demanding in terms of their degree of transparency, reliability and their ability

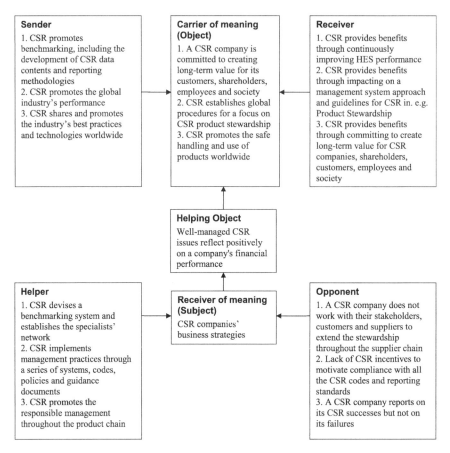

Fig. 3 Actantial model of the CSR application in relation to CSR companies' business strategies

to be audited. For many companies, intangible assets (such as brand, corporate reputation, intellectual capital, etc.) have a greater financial value than fixed assets. Any damage to intangible assets could significantly affect the overall value of the company. The understanding of what these image risks might be and conversely what opportunities could be exploited is a growing area of image risk (and possibility) management.

Hegemony: To be effective, a CSR strategy has to enjoy management commitment to ensure that the strategy is not being implemented in an ad hoc fashion. This can be achieved by establishing management processes so that risks can be identified and managed, the brand is improved and reputation is protected. The creation of a good image in public policies enables a company to translate its corporate values into commitments and responsibilities, and define its objectives and targets. The results of the economic risk assessment in turn could affect the price of financial loans, when these considerations have the so-called price tag.

Recontextualization: When CSR has been correctly implemented by a company, it can expect to receive funding from sources, which support an ethically sustainable and evolving activity, as well as foreseeing and predicting the sustainable future. Financiers will find it easier to grant financial resources, if they do not have any doubts regarding legal action being taken against the company in which they are interested in financing. The financier set-oriented funding objectives are important and the further regular reporting on CSR development is particularly important also in the value of collateral securities. The number of ethical investors continues to grow, so the status of CSR improves. When establishing the CSR image, the annual reports, CSR reports, company websites with CSR results, etc., are important. Furthermore, companies may also receive a wide range of direct queries from the various stakeholders. Another reporting of great importance involves on the social capital, knowledge management and the staff financial statement. A CSR report can be integrated into the annual report and be displayed also on web sites. Most SMEs are not even used to describing their activities in CSR terms, even though their business may involve considerable CSR elements. The major challenge by SMEs is to make visible the existing CSR operations.

Operationalization: When one looks at today's normal stock market, CSR does seem not seen to carry much significance. The only clear benefit of CSR is that the company can concretize a number of the responsibility parameters, and thus establish a wider investor base. Ethical funding will be not directed towards companies which pursue irresponsible policies. Fund managers will actively seek to invest in companies whose products or services are of long-term benefit to the communities in which they operate and contribute to a better environment. Socially responsible investment is growing faster than the general market, and this trend is expected to continue, although perhaps at a slower rate over the medium term. Profit expectations need to be extended longer than the quarterly terms.

Figure 4 presents the semiotic actantial model developed by Greimas (1987, 1990), connecting the theory and the discourse data of CSR in relation to financiers and shareholders.

5.3 Customers and Consumers

Emergence: Corporate values provide an understanding of what the values to which the company adheres, the activities for which it is prepared to be take responsibility and designates its future goals and objectives. They often cover terms such as the vision, the corporate purpose, the mission statement or goals. In the partnership, when one introduces the environmental aspects, the innovative CSR company aims to meet the needs of its customers by providing efficient, high-quality solutions that improve the productivity, end-product quality and eco-efficiency of its customers' industrial processes.

Hegemony: At the same time, innovative environmental solutions as part of the product and services portfolio can represent new business opportunities.

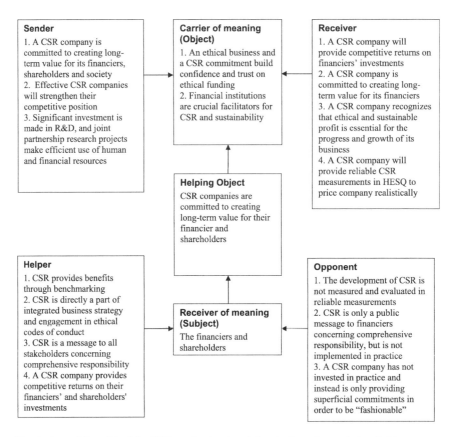

Fig. 4 Actantial model of the CSR application in relation to financiers and shareholders

The solutions may also enhance the competitiveness of the customers. The customers ultimately decide on their own choices, i.e. whether they will reward responsible behaviour or not and whether they are willing to pay a premium for this value. In this regard, there is also a conflict between consumers' values and their concrete purchasing decision.

Recontextualization: Consumers play an important part when it is a question of creating incentives for responsible production and responsible business practice. The requirements for sustainable development data have become progressively more demanding in the extent to which they include transparency, reliability and their ability to be audited. A company's capacity to manage and report on their sustainable development performance plays an integral role in the management of their risk and enhancement of operational performance. Consumers expect that companies need to display their commitment most directly by providing good jobs to people in the community.

Operationalization: A company needs to concretize CSR and report it reliably to its customers and consumers. This succeeds with vast majority of its customers, when they share the same values and ethical principles. Consumers now expect companies to be more socially responsible. They expect CSR to extend beyond making financial contributions to some charity. Corporations should not expect that consumers will obtain their information from company-generated sources alone. A great number of consumers will check up on any claims. Companies also need to look for new ways to inform consumers about their social responsible actions, which are increasingly perceived as more jobs, environmentally responsible activity and local community interaction. The key elements in good media communication are transparency, comparability and reliable indicators. The Internet is of great importance today, as CSR reports and other information are easily accessible to ordinary consumers and other stakeholders. Consumers, empowered by the Internet, are acquiring new levels of corporate citizenship. Discussions on CSR on the web are used to seek information about whether or not companies are acting in a socially responsible manner. The numbers of consumers using the Internet to research companies through independent CSR sources are growing. Consumers are using the Internet to spread information about CSR. Consumers forward e-mails, and launch CSR campaigns to advocate for some ethical position or cause.

Figure 5 presents the semiotic actantial model developed by Greimas (1987, 1990), connecting the theory and the discourse data of CSR in relation to customers and consumers.

5.4 Employees of the Companies

Emergence: The CSR combines the company's own intentions, as expressed in its values and code of conduct in accordance with the expectations of its stakeholders. In the commitment to CSR, the company expresses its commitment to the promotion of sustainable development and its three pillars: economic, environmental and social sustainability. The company's code of conduct sets out common values and ethical principles which are being applied, emphasizing reliability, openness and fairness.

Hegemony: Code of conduct deals with issues such as clear and transparent operational and decision-making structures, fair cooperation with employees, customers, suppliers, colleagues and other stakeholders, and careful handling of confidential information. Paying attention to staff welfare, developing staff competencies and the promoting of the intellectual capital can contribute to a company's ability to attract a skilled labour force.

Recontextualization: A company with good CSR policies will be able to generate commitment among its staff. CSR will become an important factor in competing for the best employees in the future. Including objectives and targets in the performance appraisal system promotes the integration of corporate responsibility issues throughout an organization. Similarly, training and development can help increase the employees' appreciation of the concept of CSR and its significance for the

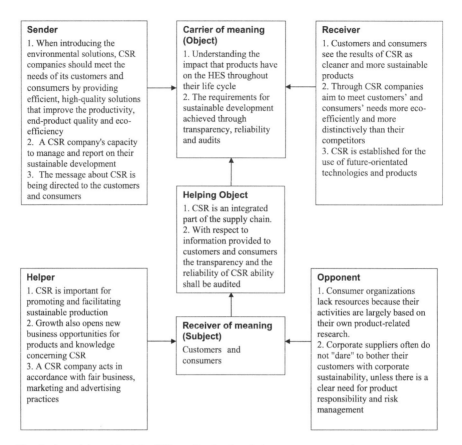

Fig. 5 Actantial model of the CSR application in relation to customers and consumers

company; this in turn should mean that decisions will be made with greater knowledge of the risks and opportunities for the business. The occupational safety and health (OSH) of employees and how this is affected by its operations which are core components of the CSR company. OHS practices can also be based on the CSR company's ethical values, openness, trust and innovativeness and its best practices of social responsibility.

Operationalization: The company complies with international, national and local rules, regulations and agreements. The company's OSH activities are directed by the principles of continuous improvement with an emphasis on quality and know-how about sustainability. The OSH issues impacted by its operations are essential to the CSR company. The company complies with international, national and local rules, regulations and agreements. The company's OSH activities are governed by the principles of continuous improvement with an emphasis quality, innovations, know-how and supporting lifelong learning in the employees.

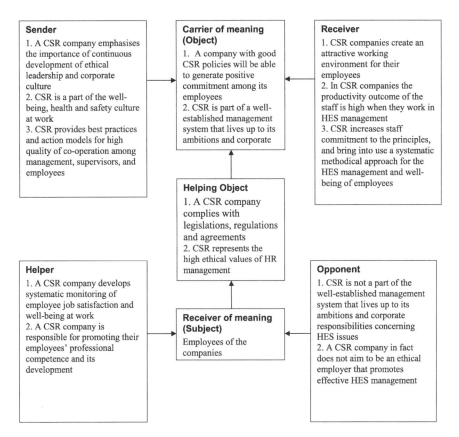

Fig. 6 Actantial model of the CSR application in relation to employees of the companies

Figure 6 presents the semiotic actantial model developed by Greimas (1987, 1990), connecting the theory and the discourse data of CSR in relation to employees of the companies.

5.5 Communities and Authorities

Emergence: In addition to the role of the legislator and the employer, public policy may also take on the responsibility of the client in the case of public procurement. Public procurement is often a highly regulated process with a number of preconditions and some of these are clearly CSR issues, which need to be taken into account. In fact, CSR may also often be defined in the legislation. The problems for international companies may be that there are different types of the legislation in several countries which all need to be included in the company's CSR policy.

The NGO stakeholders can be a gateway to ensure that community and authorities will become involved in promoting better corporate citizenship. One must hope that ethics and social responsibility will have a positive impact on the success of an organization, especially non-governmental stakeholders will make ethical judgments that are likely to influence any purchases. NGOs may also trigger consumer boycotts. There are numerous examples where these have been effective. Nowadays companies are fearful of this kind of negative publicity that they usually correct working practices if it is at all financially feasible.

Hegemony: Community and authorities expect companies not only to obey their legal obligations but also to be socially responsible. They expect CSR to go beyond simply a making financial contribution to some good cause. Community and authorities can encourage enterprises to become better corporate citizens. The governmental and local authorities can also promote the development of a company's CSR policies through initiating a variety of educational projects. Opportunities for partnership in joint projects for example promoting cooperation between companies and the governmental and local authorities may be advantageous.

The NGO stakeholders believe that they must have the opportunity to select or reject certain products or suppliers, even multinational companies depending on whether they are perceived to be acting more or less responsibly in terms of CSR. Some NGO organizations have also been involved in partnership-type joint projects with certain companies. Often a socially aware CSR with good CSR policies dovetail with the sustainable development championed by many NGOs. The task of associations which focuses on environmental protection, as well as consumers' organizations, is often to act as 'neutral' controllers, in relation to the actual CSR achievements.

Recontextualization: The local community has often a particularly important role to play, because many of the company's employees are also members of the local community. A very important issue is to take account of authority stakeholders' expectations and to support interactions, e.g. in situations environment cooperation, it is important to act, in a positive way in order to meet the authority stakeholders' expectations. CSR is 'the extraordinary good', which companies create in addition to fulfilling their statutory and legislation obligations. At present, the development of CSR can be considered to be advancing as a part of best practices partly being driven by market forces.

If CSR is only a platitude, the so-called 'green washing' as footnote in a company's action plan, then it is likely that NGO associations will react against such companies and publicize their hypocrisy. One reason for the many CSR brands is the growing interest of consumers in buying sustainable products and services. Nonetheless, consumers' enthusiasm, awareness and willingness to pay have not yet been converted into a situation where the majority of the consumers are ready to open their wallets to pay for more sustainable choices.

Operationalization: The main focus is on the CSR alliance promotion. In the future, one can anticipate that there will be ever more stringent regulations gradually encompassing all areas, in which a company operates including surveillance, management culture, control, as well as reporting. The ISO Standard 26000:2010 focuses the guidance type of social responsibility on innovation. Innovation is the key to creating win–win situations. There are four broad areas where responsible businesses may encounter a CSR risk: operational risks, regulatory risks, reputation risks and sustainability/strategic risks. Companies need to establish reliable ways to identify important ethical business-based issues, and to incorporate these into their overall risk management systems. Many companies use suites of policies to map the importance of specific governmental regulations, community-impacts, as well as environment, workplace and marketplace issues, and this can help them control risk within those areas. It is clear that these policies must be dynamic and regularly reviewed, so that they reflect the current priorities not only of the business but also of society as a whole.

NGO associations often are the first groups to identify a company's accountability priorities and to trigger changes in the way the companies operate. The NGO associations have different impacts, and the relevance of the publicity generated their activities will vary from company to company.

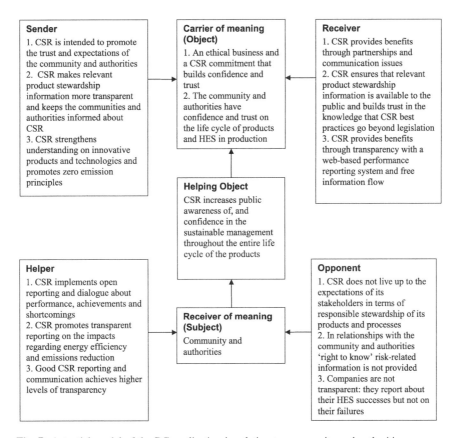

Fig. 7 Actantial model of the RC application in relation to community and authorities

Nonetheless, it is important that all businesses view seriously the specific problems highlighted by NGO associations. It is important that there is a continuous dialogue between the company and these associations. Generally speaking, the NGO associations tend to set very ambitious targets for companies with respect to CSR.

Figure 7 presents the semiotic actantial model developed by Greimas (1987, 1990), connecting the theory and the discourse data of CSR in relation to community and authorities.

6 Discussion

6.1 Comparing the Results with Other Studies

The results of this study are similar with the findings of Windell (2006, p. 17) that the ideas and arguments of CSR are constructed in a translation processes involving a diverse groups of actor and, in particular, that the social partners are not passive adopters of ideas; instead they play an important role in translation processes (Windell 2006, p. 17).

The results of this study support the findings of Windell (2006, p. 41), i.e. if one wishes to understand how and why ideas about CSR are formed, created and recreated one needs to address how social partners circulate ideas. The concept of translation conceptualizes the circulation and construction of ideas in CSR as processes in which concepts are materialized and endowed with meaning. In addition, the results support the conclusions of Salzmann et al. (2006, p. 3), i.e. the interactions between stakeholders and companies in CSR-related issues are characterized by the fact that stakeholders contribute something to the company and in exchange expect that their interests will be satisfied by some kind of return.

The results of this study are similar to the findings of other investigators (e.g. Salzmann et al. 2006, pp. 3–4) that the fundamental organizing principles of CSR are as follows: (1) the companies' economic responsibility to generate profits; (2) the legal responsibility of a business to be in compliance with existing regulation; (3) the ethical responsibility refers to the fulfilment of society's expectations; and (4) philanthropic actions result in social capital benefit of human resources. The results support the findings of Ramus and Oppegaard (2007, p. 5) that the integration of value-driven activities with implementation-focused processes will be achieved differentially in each case, depending on a series of characteristics unique to each organization. However, Ramus and Montiel (2005, p. 378) stated that similar policies seem to underlie different kinds of sustainability management systems in different companies.

The results are similar to the works of Ramus and Oppegaard (2006, p. 12) who claimed the top management has to make an appropriate strategic response to actual real-life situations. The results support the results of others that the financial markets now demand that companies exercise transparency in their operations (e.g. Dobers and Wolff 2000) and today the large financial institutions can exert a major influence on how a company must operate within society (Steger 2006, p. 4). Furthermore, today, one does see the appearance of socially responsible investing ratings although these are still not very insignificant (Steger 2006, p. 4). The result support the findings of others (e.g. Dobers and Wolff 2000) that the consumers seem to express positive environmental attitudes, but this only is reflected seldom in their behaviour (Dobers and Wolff 2000).

6.2 Discussion on the Generalizability of the CSR Discourse Analysis

The standardized open-ended online questionnaire consisted of a set of CSR questions carefully worded and arranged with the intention of taking the respondents through the same sequence and asking each respondent the same questions phrased with essentially the same words. The questions were formulated in advance and provided in the sequence in which they were to be answered. The basic strength of the standardized open-ended interview is that the respondents answer the same questions, thus increasing comparability of responses. The advantage of this questionnaire is that it makes the answers provided by a number of different individuals in a more systematic and comprehensive manner limiting the issues to be handled in the interview. The weakness of this approach is that it does not permit the interviewer to pursue topics or issues that were not anticipated when the questionnaire instrument was elaborated. In addition, a number of questions were included in the framework allowed which the respondent the possibility to respond in greater depth. Thus, the strength of this latter approach is that this part of the questionnaire is flexible and highly responsive to individual differences, situational changes and emerging new information.

In the CSR process, concepts are also described meaningfully since they can be presented in accordance with the existing institutional context. Hence, by evaluating a broad range of actor groups, one can potentially clarify how translation processes proceed as a part of an interaction between actors that intuitively seek to translate their ideas in accordance with their own interests and working practices (Windell 2006, p. 41).

Other research methods could be applied to accompany with DA, e.g. the methods developed by other groups (e.g. Leca and Naccache 2006; Vincent 2008). Leca and Naccache (2006) developed a model with which to analyse institutional pressures in an attempt explain actors' actions and behaviours and to consider simultaneously the influence of both actors' actions and the structures in which they are embedded. Vincent (2008) created a critical realistic model to wish which to investigate interorganizational relationships and networks and which can identify the impact of social processes and how these are shaped by the behavioural orientations of the actors. This realistic but critical approach provides a perspective on agency and social processes that facilitate analyses of the behavioural orientations of actors and how these combine in order influence on processes and outcomes.

Qualitative discourse research (Fairclough 2003; 2005) represents a naturalistic approach that seeks to understand RC phenomena in context-specific settings. This is related to the concept of good quality research when reliability has the purpose of 'generating understanding' (Stenbacka 2001; Lincoln and Guba 1985, p. 317; Healy and Perry 2000; Hoepfl 1997; Patton 2002; Potter 1996). If the validity or trustworthiness of discourse research into CSR can be maximized, then one can obtain more 'defensible results' (see Johnson 1997) and this may lead to better generalizability which is one of the concepts proposed by Stenbacka (2001) as the structure for documenting high-quality qualitative research. The concept of validity is described by a wide range of terms in qualitative studies. Creswell and Miller (2000) suggested that the validity is affected by the investigator's perception of validity in the study and his/her choice of paradigm assumptions. In qualitative research, the principle of discovering truth through measures of reliability and validity is replaced by the concept of trustworthiness, which is 'defensible' (Johnson 1997) and establishing confidence also in the findings of the discourses in this case about CSR (see Lincoln and Guba 1985). Therefore, the results of this discourse research on CSR have adequate generalizability and trustworthiness.

7 Conclusions and Recommendations

In a world of complexity and change, the commercial enterprises are expected to be able to cope with much greater diversity of CSR problems. They have to continue to ensure that their organizational CSR processes are efficient and that they are served by the latest developments in technology. Within CSR, the top management has to make an appropriate strategic response to actual real-life situations. A central consideration for business organizations must be how well private CSR initiatives reflect and reinforce government agreements on labour, social and environmental standards. The financial markets now demand that companies ensure transparency in their operations and today the large financial institutions can exert a major influence on how a company must operate within society. The consumers seem to express positive environmental attitudes relating to CSR, but this is seldom reflected in their behaviour. Business and society represent twin entities, i.e. (1) business firms are social institutions and hence (2) these institutions must earn their legitimacy from multiple sectors of society.

The CSR is seen as that part of the overall strategic management of the organization that focuses specifically on the social and ethical issues embedded into the functioning and decision processes of the firm. The CSR processes occur in the context of both organizational (internal) and environmental (external) forces. In the light of these challenges, a conceptual framework of CSR can be devised and its different aspects can be analysed from perspective of stakeholder-specific factors that influence the effectiveness of CSR. By engaging in CSR activities, companies not only can generate favourable stakeholder attitudes and better support behaviours but also, in the long run, build a good corporate image, strengthen stakeholder–company relationships, and enhance stakeholders' advocacy behaviours. However, stakeholders' low awareness of companies' CSR activities remain as a critical impediment in companies' attempts to maximize business benefits from their CSR activities.

Acknowledgments The author are deeply grateful to the following people who gave generously of their time and produced their discursion data of this study: Hans Grahn (Senior CSR Specialist) in the Confederation of Finnish Industries EK, Jyrki Helin (Responsible Research Specialist) in the Central Organization of Finnish Trade Unions SAK and Juha Beurling (Project Manager of Ethical Consumers) in the Finnish Consumers' Association.

References

Bansal P, Roth K (2000) Why companies go green: a model of ecological responsiveness. Acad Manag Rev 43(4):717–736
Business in the Community (2008) What's in it for you—why businesses should invest in their local community. Publication of the Business in the Community. http://www.bitc.org.uk/resources/publications/index.html. Accessed 4 June 2014
Carroll AB (2000a) Ethical challenges for business in the new millennium: CSR and models of management morality. Bus Ethics Q 10(1):133–142
Carroll AB (2000b) A commentary and an overview of key questions on corporate social performance measurement. Bus Soc 39(4):466–478
Catasus B, Lundgren M, Rynnel H (1997) Environmental managers' views on environmental work in a business context. Bus Strategy Environ 6(4):197–205
Cavanagh GF, Moberg DJ, Velasquez MG (1995) Making business ethics practical. Bus Ethics Q 5(3):399–418
Christmann P (2000) Effects of "best practices" of environmental management on cost advantage: the role of complementary assets. Acad Manag J 43(4):663–680
Clavera E, López MD, Molina JF, Taria JJ (2007) The link between 'green' and economic success: environmental management as the crucial trigger between environmental and economic performance. J Environ Manage 84(4):606–619
Commission of the European Communities (2002) The communication on CSR (2.7.2002) com ((2002) 347 final) concerning a business contribution to sustainable development. http://eur-lex.europa.eu/LexUriServ/LexUriServ.do?uri=COM:2002:0347:FIN:EN:pdf. Accessed 4 June 2014
Commission of the European Communities (2006) The new communication on CSR (22.3.2006) com ((2006) 136 final) concerning the implementing the partnership for growth and jobs, and making Europe a pole of excellence on CSR. http://eur-lex.europa.eu/LexUriServ/LexUriServ.do?uri=COM:2006:0136:FIN:EN:PDF. Accessed 4 June 2014
Confederation of Finnish Industries EK (2014a) Economy and trade: the ageing population poses a challenge for the national economy. http://www.ek.fi/www/en/economy_trade/index.php. Accessed 4 June 2014
Confederation of Finnish Industries EK (2014b) Interaction is essential in the innovation system Confederation of Finnish Industries EK. http://www.ek.fi/www/en/innovation_education/index.php. Accessed 4 June 2014
Confederation of Finnish Industries EK (2014c) Corporate responsibility—practices in Finnish business. http://www.ek.fi/. Accessed 4 June 2014
Crane A (1999) Are you ethical? Please think yes or no on researching ethics in business organizations. J Bus Ethics 20(3):237–248
Crane A (2000) Corporate greening as a moralization. Organ Stud 21(4):673–696
Creswell JW, Miller DL (2000) Determining validity in qualitative inquiry. Theory Pract 39 (3):124–131

Czarniawska B, Joerges B (1996) Travels of ideas. In: Czarniawska B, Sevon G (eds) Translating organizational change. Walter de Gruyter, Berlin, pp 13–47

Cunliffe A (2008) Orientations to social constructionism: relationally responsive social constructionism and its implications for knowledge and learning. Manag Learn 39(2):123–139

Deakin S, Hobbs R (2007) False dawn for CSR? Shifts in regulatory policy and the response of the corporate and financial sectors in Britain. Corp Gov 15(1):68–76

Dobers P, Wolff R (2000) Competing with 'soft' issues—from managing the environment to sustainable business strategies. Bus Strategy Environ 9(1):143–150

Elkington J (2001) The Chrysalis economy. Join citizen CEOs and corporations can fuse values and value creation. Capstone Publishing Ltd, Oxford

European Parliament Resolution (2006) Session document final, 20.12.2006. Report on corporate social responsibility: a new partnership (2006/2133(INI)). EU Committee on Employment and Social Affairs. http://www.europarl.europa.eu/sides/getDoc.do?type=REPORT&reference=A6-2006-0471&language=EN&mode=XML. Accessed 4 June 2014

Fairclough N (1995) Critical discourse analysis. Longman, London

Fairclough N (1996) Discourse and social change. Polity Press, Cambridge

Fairclough N (2000) New labour, new language. Routledge, London

Fairclough N (2003) Analysing discourse: textual analysis for social research. Routledge, London

Fairclough N (2005) Critical discourse analysis, organizational discourse, and organizational change. Organ Stud 26(2):915–939

Fairhurst G, Grant D (2010) The social construction of leadership: a sailing guide. Manag Commun Q 24(2):171–210

Fineman S (1996) Emotional subtexts in corporate greening. Organ Stud 17(3):479–500

Fineman S (1997) Constructing the green manager. Br J Manag 8(1):31–38

Finnish National Commission on Sustainable Development (2006) Towards sustainable choices: a nationally and globally sustainable Finland. The national strategy for sustainable development. Prime Minister's Office Publications 7. Edita, Helsinki. http://www.environment.fi/sustainable-development. Accessed 4 June 2014

Finnish Trade Union Movement (2014) What every employee should know? SAK, STTK and Akava http://www.sak.fi/english/whatsnew.jsp?location1=1&sl2=1&lang=en. Accessed 4 June 2014

Finnish Strategy (2011) Socially sustainable Finland 2020. Strategy for social and health policy. Ministry of Social Affairs and Health, Helsinki. http://www.stm.fi/c/document_library/get_file?folderId=2765155&name=DLFE-15361.pdf. Accessed 4 June 2014

Fisher C, Lovell A (2006) Business ethics and values: individual, corporate and international perspectives, 2nd edn. Pearson Education Limited, Harlow

French PA (1993) The corporation as a moral person. In: White TI (ed) Business ethics: a philosophical reader. Macmillan, New York, pp 228–235

Freeman RE, Liedtka J (1997) Stakeholder capitalism and the value chain. Eur Manag J 15 (3):286–296

Friedman M (1970) The social responsibility of business is to increase its profits. New York Times Magazine. September 13

Greimas AJ (1987) On meaning: selected writings in semiotic theory. University of Minnesota Press, Minneapolis

Greimas AJ (1990) The social sciences: a semiotic view. University of Minnesota Press, Minneapolis

Hart SL (1995) A natural-resource-based view of the firm. Acad Manag Rev 20(4):986–1014

Hart SL, Ahuja G (1996) Does it pay to be green? An empirical examination of the relationship between emission reduction and firm performance. Bus Strategy Environ 5(1):30–37

Healy M, Perry C (2000) Comprehensive criteria to judge validity and reliability of qualitative research within the realism paradigm. Qual Mark Res 3(3):118–126

Hébert L (2011) Tools for text and image analysis: an introduction to applied semiotics. Université du Québec à Rimouski, Québec. http://www.signosemio.com/documents/Louis-Hebert-Tools-for-Texts-and-Images.pdf. Accessed 21 July 2014

Hoepfl MC (1997) Choosing qualitative research: a primer for technology education researchers. J Technol Educ 9(1):47–63

ILO (2006) Tripartite declaration of principles concerning multinational enterprises and social policy. International Labour Office, Geneva. http://www.ilo.org/public/english/employment/multi/download/declaration(2006).pdf. Accessed 4 June 2014

ISO Standard 26000 (2010) Guidance on social responsibility. ISO (International Organization for Standardization), Geneva

Jackson MC (2003) Systems thinking: creative holism for managers. John Wiley & Sons Ltd, West Sussex. http://webcourses.ir/dl/Systems%20Thinking.pdf. Accessed 4 June 2014

Johnson BR (1997) Examining the validity structure of qualitative research. Education 118(3):282–292

Leca B, Naccache P (2006) A critical realist approach to institutional entrepreneurship. Organization 13(5):627–651

Lincoln YS, Guba EG (1985) Naturalistic inquiry. Sage, Beverly Hills

Lord RG, Dinh JE (2012) Aggregation processes and levels of analysis as organizing structures for leadership theory. In: Day DV, Antonakis J (eds) The nature of leadership, 2nd edn. Sage Publications Inc., Thousand Oaks, pp 29–65

Moir L (2001) What do we mean by corporate social responsibility. Corp Gov 1(2):16–22

OECD-ILO (2008) Conference on corporate social responsibility. Overview of selected initiatives and instruments relevant to corporate social responsibility. Promoting responsible business conduct in a globalising economy 23–24 June 2008. OECD Conference Centre, Paris, France. http://www.oecd.org/dataoecd/18/56/40889288.pdf. Accessed 4 June 2014

OECD (2005) The OECD guidelines for multinational enterprises. http://www.oecd.org/dataoecd/56/36/1922428.pdf. Accessed 4 June 2014

Orlitzky M, Schmidt FL, Rynes SL (2003) Corporate social and financial performance: a meta-analysis. Organ Stud 24(3):403–441

Patton MQ (2002) Qualitative evaluation and research methods, 3rd edn. Sage Publications Inc., Thousand Oaks

Pirson M, Malhotra D (2008) Unconventional insights for managing stakeholder trust. MIT Sloan Manag Rev. http://www.hbs.edu/research/pdf/08-057.pdf. Accessed 4 June 2014

Potter J (1996) Representing reality: discourse, rhetoric and social construction. Sage, London

Potter J (1997) Discourse analysis as a way of analysing naturally occurring talk. In: Silverman D (ed) Qualitative research: theory, method and practice. Sage Publications, London, pp 144–160

Potter J (2000) Post cognitivist psychology. Theory Psychol 10(2):31–37

Ramus C (2005) Context and values: defining a research agenda for studying employee environmental motivation in business organizations. In: Aragon-Correa A, Sharma S (eds) Research in corporate sustainability. Edward Elgar Publishing, Cheltenham, pp 71–95

Ramus CA, Killmer ABC (2007) Corporate greening through prosocial extrarole behaviours—a conceptual framework for employee motivation. Bus Strategy Environ 16(8):554–570

Ramus C, Montiel I (2005) When are corporate environmental policies a form of "greenwashing"? Bus Soc 44(4):377–414

Ramus CA, Oppegaard K (2006) Shifting paradigms in sustainability management: instrumental and value-driven orientations. International Institute for Management Development (IMD) Publication No. 1. http://www.imd.ch/research/publications/upload/Ramus_Oppegaard_WP_(2006)_01_Level_1.pdf. Accessed 4 June 2014

Ramus CA, Oppegaard K (2007) Integrating compliance-based and commitment-based approaches in corporate sustainability management. International Institute for Management Development (IMD) Publication No. 4. http://www.imd.ch/research/publications/upload/Ramus_Oppegaard_WP_(2007)_04_Level_1.pdf. Accessed 4 June 2014

Salzmann O, Steger U, Ionescu-Somers A (2005) Quantifying economic effects of corporate sustainability initiatives—activities and drivers. International Institute for Management Development (IMD) Publication No. 28. http://www.imd.ch/research/centers/csm/upload/Quantifying%20Economic%20Effects%20BCS%20WP%20(2005)-28%20.pdf. Accessed 4 June 2014

Salzmann O, Steger U, Ionescu-Somers A, Baptist F (2006) Inside the mind of stakeholders—are they driving corporate sustainability? International Institute for Management Development (IMD) Working Paper, October. http://www.imd.ch/research/publications/upload/CSM_Salzmann_Steger_Ionescu_Somers_Baptist_WP_(2006)_22.pdf. Accessed 4 June 2014

Sharma S (2000) Managerial interpretations and organizational context as predictors of corporate choice of environmental strategy. Acad Manag J 43(4):681–697

Sharma S, Vredenburg H (1998) Proactive corporate environmental strategy and the development of competitively valuable organizational capabilities. Strateg Manag J 19(1):729–753

Senge PM (1994) The fifth discipline: the art and practice of the learning organization, 2nd edn. Bantam Doubleday/Currency Publishing Group, New York

Steger U (2006) Inside the mind of the stakeholder. A preliminary excerpt from U. Steger (Ed.) (2006): inside the mind of the stakeholder—the hype behind stakeholder pressure. Palgrave Macmillan, Basingstoke. http://www.imd.ch/research/centers/csm/upload/Executive%20summary%20-%20Stakeholder%20book.pdf. Accessed 4 June 2014

Stenbacka C (2001) Qualitative research requires quality concepts of its own. Manag Decis 39(7):551–555

Strauss AL, Corbin J (1998) Basics of qualitative research, 2nd edn. Sage, London

Tsoukas H, Chia R (2002) On organizational becoming: rethinking organizational change. Organ Sci 13(5):567–585

United Nations (2007) Business and human rights: mapping international standards of responsibility and accountability for corporate acts. Report of the Special Representative of the Secretary-General. http://www.business-humanrights.org/Documents/SRSG-report-Human-Rights-Council-19-Feb-2007.pdf. Accessed 4 June 2014

van Dijk TA (1995) Discourse semantics and ideology. Discourse Soc 6(2):243–289

van Dijk TA (2001) Critical discourse analysis. In: Schiffrin D, Tannen D, Hamilton HE (eds) The handbook of discourse analysis. Blackwell Publishing, Massachussets and Oxford, pp 363–415

Velasquez MG (1992) Business ethics: concepts and case. Prentice Hall, Englewood Cliffs

Vincent S (2008) A transmutation theory of inter-organizational exchange relations and networks: applying critical realism to analysis of collective agency. Hum Relat 61(6):875–899

Wagner M (2005) How to reconcile environmental and economic performance to improve corporate sustainability: corporate environmental strategies in the European paper industry. J Environ Manag 76(2):105–118

Welford R (2000) Corporate environmental management: towards sustainable development. Earthscan Publications, London

Wetherell M (1998) Positioning and interpretative repertoires: conversation analysis and post-structuralism in dialogue. Discourse Soc 9(1):387–412

Wetherell M, Taylor S, Yates SJ (2001) Discourse theory and practice: a reader. Sage, London

Windell K (2006) Corporate social responsibility under construction: ideas, translations, and institutional change. Upsala Universitet, Företagsekonomiska institutionen. Department of Business Studies, Doctoral thesis No. 123. Universitettryckeriet, Ekonomikum, Uppsala

Wood DJ (1991) Corporate social performance revisited. Acad Manag Rev 16(4):691–718

Chapter 14
A Case Study About the Discourses of the Finnish Corporations Concerned with Sustainable Development

Toivo Niskanen

Abstract The aim of the present study was to explore (1) the different kinds of practices implemented by Finnish corporations with respect to the "corporate sustainability" and how (2) "corporate sustainability" is applied in their business practices. A further aim was to describe and explain the cross-sectional association mechanisms, context and outcome patterns of the impact of "corporate sustainability". The values of an organization are the criteria on which decisions about priorities are made. Some corporate values are ethical such as those that guide the decisions and especially the processes. Within the framework of "corporate sustainability" values have a broader meaning. An organization's values are the standards by which managers and employees make prioritization decisions. The "corporate sustainability" can be applied in different ways: (1) "Corporations' business strategies" when creating long-term positive values and establishing proactive procedures; (2) "Financiers and shareholders" when building trust and achieving long-term sustainability; (3) "Customers and consumers" when accounting for potential impacts and providing transparent information; (4) "Employees of the corporations" when implementing proactive measures and ensuring good sustainability as a part of a well-established management; (5) "Communities and authorities" when establishing a confidence in the supply-chain of products, production and services and creating active and transparent processes by provision of relevant information, as well arranging meetings, public hearings and undertaking consultation procedures.

Keywords Corporate · Sustainability · Development · Discourses · Case study · Environment · Health · Safety

---

T. Niskanen (✉)
Ministry of Social Affairs and Health, Occupational Safety and Health Department,
Legal Unit, P.O. Box 33, FI-00023 Government, Helsinki, Finland
e-mail: toivo.niskanen@stm.fi

1 Introduction

Drucker (2001, p. 15) stated that "The task of management is managing the social impacts and the social responsibilities of the enterprise. Every enterprise is an organ of society and exists for the sake of society. Business is no exception. Free enterprise cannot be justified as being good for business; it can be justified only as being good for society."

Although the concept of sustainability was developed at the macro level rather than at the corporate level, it can also be considered to have a relevant corporate dimension (de Lange et al. 2012). The term came into widespread use in 1987, when the World Commission on Environment and Development (United Nations) published a report known as the 'Brundtland Report'; this report stated that 'the "corporate sustainability" seeks to meet the needs of the present without compromising the ability to meet the future generation to meet their own needs' (WCED 1987, p. 8). Although this report originally only included environmental aspects, the concept of 'the "corporate sustainability"' has since expanded to incorporate the consideration of the social dimension as being inseparable from development. In the words of the World Business Council for The "corporate sustainability" (2000, p. 2), the "corporate sustainability" "requires the integration of social, environmental, and economic considerations to make balanced judgments for the long term''.

The "corporate sustainability" (e.g. Zink 2014) or as in the present study, "corporations" ways to deal with the "corporate sustainability" means first of all to discuss the relevance of sustainability as the simultaneous pursuit of economic, ecological and social objectives (WCED 1987). The "corporate sustainability" (as a process) relies on following three basic ideas (e.g. Zink 2014): (1) The focus on human needs: 'Human beings are in the centre of concerns for the "corporate sustainability". They are entitled to a healthy life in harmony with nature' (UNCED 1992); (2) The normative claim for intra- and intergenerational fairness as it was stated in the well-known definition of the "corporate sustainability" (WCED 1987); (3) The concurrent combination of economic, ecological and social goals; these are three pillars of the "corporate sustainability", which should be considered equally (UNCED 1992).

Transferring the general definition on a corporate level leads to the concepts of the "corporate sustainability" (Zink 2014) come along with the following definition elements (e.g. Zink 2014; Dyllick and Hockerts 2002): (1) Not only economic but also social and environmental prerequisites and impacts as well as the interdependencies between them have to be taken into account; (2) Corporate sustainability requires a long-term business orientation as a basis for satisfying stakeholders' needs now and in the future; and (3) The rule to live on the income from capital, not the capital itself has to be applied for all kinds of capital: financial, natural, human and social capital.

The term 'stakeholding' refers to a concept, a principle or argument, whereas the term stakeholder refers to a specifiable person or groups of individuals (sometimes in an organized form), with clear implications for how the interests of such groups might be incorporated or represented within organizational decisions (Fisher and Lovell 2006, p. 319). Nowadays, corporate managers are becoming aware of the need to broaden their goals beyond the traditional financial expectations (de Lange et al. 2012). Furthermore, since the term sustainability has entered the business world, an ever-increasing number of firms have started to appreciate the importance of sustainability and now emphasize the social and environmental goals of their organizations. Sustainability can be defined, based on earlier definitions, as an approach to business that considers economic, environmental and social issues in balanced, holistic and long-term ways that benefit current and future generations of concerned stakeholders (World Commission on Environment and Development 1987). Furthermore, de Lange et al. (2012) stated that national culture can help to explain the variability in adoption. In this context, sustainability can be considered as a "norming" effect which occurs when national culture influences the development of legitimation pressures. Corporate responsibility, corporate governance and corporate citizenship provide the initial focus for the consideration of business ethics as seen at the organizational level (Fisher and Lovell 2006).

The simple fact that corporations are social constructs does not deny the possibility that the economic impact of such entities can develop to such an extent that society deems it necessary to place constraints, or responsibilities, upon their activities (Fisher and Lovell 2006, p. 318). The essential learning mechanism, what Schein (1996, p. 64) in his treatise on organizational culture called "cognitive redefinition" involves (1) new semantics, i.e. redefining in a formal sense what individualism means; (2) broadening perceptions to enlarge one's mental model of individualism to include collaborative behaviours as well as competitive behaviours; and (3) developing new standards of judgment and evaluation so that competitive behaviour can be viewed in a more negative way while collaborative behaviour is viewed as being more positive. It has been stated that the most efficient and economic form of corporate governance is a relationship built upon trust, buttressed by the requisite levels of accountability and transparency (Fisher and Lovell 2006, p. 321). Furthermore, nowadays national policies encompass norms and values incorporating the principles of sustainability and enterprises need to adhere to sustainability-relevant practices; these altitudes, i.e. increase the likelihood of adoption of sustainability good practices. The institutional pressures on corporations may appear at local, national and international levels (de Lange et al. 2012).

The sustainability performance and improvements of corporations may be encouraged through reporting and monitoring; an important issue has been whether voluntary or regulatory approaches are more effective in changing enterprises' behaviour in a sustainable direction (de Lange et al. 2012). There are several competencies which can act to make the "corporate sustainability" an integral part of business practice, i.e. (applied from Fisher and Lovell 2006, p. 402): (1) Understanding society and business's roles and obligations within society; (2) Building capacity within an organization to work effectively in a responsible manner; (3) Questioning the practices 'business as usual', (4) Stakeholder relations,

(5) Strategic view and ensuring that social and environmental concerns are considered in broad decision-making and (6) Harnessing diversity.

There is no single road leading to "corporate sustainability" from the many approaches each corporation should choose. The best opportunities and options match the corporation's aims and intentions which can be aligned with the corporation's strategy, as a response to the conditions in which it operates (applied from van Marrewijk 2003). The concept of corporate social responsibility has been investigated extensively. Some authors (e.g. Garriga and Melé 2004) have identified social and ethical resources and capabilities which can be a source of competitive advantage, such as the process of moral decision-making, the process of perception, deliberation and responsiveness or capacity of adaptation and the development of proper relationships with the primary stakeholders: employees, customers, suppliers and communities. In this respect, sustainability would represent a set of benchmark indicators that can be used to assess whether management and employees in the corporation (applied from Fisher and Lovell 2006, p. 402) adhering to the following concept: (1) awareness of sustainability, (2) understanding the issues surrounding sustainability, (3) applying these competencies at work, (4) integrating the competencies into the culture of the corporation and (5) providing leadership on sustainability across the organization.

Finding the balance between market growth and capacity expansion is a dynamic problem in the "corporate sustainability". Long-term success in the "corporate sustainability" depends on "the process whereby management change their traditional mental models and systems thinking in their corporation and their views about their markets, and their competitors" (applied from Senge 1994, p. 188). Fisher and Lovell (2006, p. 383) have listed eight roles for corporate codes: (1) Damage limitation, (2) Guidance, (3) Regulation. (4) Discipline and role of a code as a benchmark, (5) Information about standards of behaviour, (6) Proclamation and the role of codes of conduct, (7) Negotiation and (8) Internal procedures for handling ethical concerns.

Nowadays, many consider "corporate sustainability" and "corporate social responsibility (CSR)" as synonyms. The conclusions of van Marrewijk (2003) recommend that a minor but essential distinction should be maintained in order to associate CSR with the communal aspect of people and organizations and "corporate sustainability" with the agency principle. Therefore, CSR relates to phenomena such as transparency, stakeholder dialogue and sustainability reporting, while sustainability focuses on value creation, environmental management, environmental friendly production systems, human capital management, etc. (van Marrewijk 2003). In general, corporate sustainability and CSR refer to corporation activities—voluntary by definition—demonstrating that the enterprise incorporates social and environmental concerns into its business operations and its interactions with stakeholders. This is the broad—although some would say "vague"—definition of corporate sustainability and CSR (van Marrewijk 2003).

The results of van Marrewijk (2003) differentiate the definition of "corporate sustainability" into five interpretations measures with ambition levels of corporate sustainability. In addition, there may be different motives for choosing a particular approach (van Marrewijk 2003):

(1) Compliance-driven: corporate sustainability at this level consists of providing welfare to society, within the limits of regulations from the appropriate authorities. In addition, organizations might respond to charity and stewardship considerations. The motivation can be viewed as a duty and obligation, or simply as good behaviour;
(2) Profit-driven: corporate sustainability at this level consists of the integration of social, ethical and ecological aspects into business operations and decision-making, provided it contributes to the financial bottom line. The motivation is a business case: corporate sustainability will be promoted if it is viewed as profitable, for example because of an improved reputation for the enterprise in its activity sectors in various markets (e.g. customers, employees, shareholders).
(3) Caring: corporate sustainability consists of balancing economic, social and ecological concerns, which are all three important in themselves. The initiatives go beyond those demanded by legal compliance and beyond profit considerations. The motivation for corporate sustainability is that human potential and social responsibility are important as such.
(4) Synergistic: corporate sustainability consists of a search for well-balanced, functional solutions creating value in the economic, social and ecological realms of corporate performance, in a synergistic, win-together approach with all relevant stakeholders. The motivation is that sustainability is important in itself, especially because it is recognized as being the inevitable direction that progress takes.
(5) Holistic: corporate sustainability is fully integrated and embedded in every aspect of the organization, aimed at contributing to the quality and continuation of life of every being and entity, now and in the future. The motivation is that sustainability is the only alternative since all beings and phenomena are mutually interdependent.

Globally, the UN Global Compact's (2014) ten principles in the areas of human rights, labour, the environment and anti-corruption enjoy universal consensus and are derived from: (1) The Universal Declaration of Human Rights, (2) The International Labour Organization's (ILO) Declaration on Fundamental Principles and Rights at Work, (3) The Rio Declaration on Environment and Development, (4) The United Nations Convention Against Corruption and (5) The OECD (2011) Guidelines for Multinational Enterprises.

The current requirements of the Global Reporting Initiative (GRI) require organizations to report on the corporate environmental, social and economic performance information, in essence to estimate the corporation's sustainability performance. The GRI should include the following framework documents, which all together can be viewed as a kind of 'Sustainability Report': (1) Sustainability reporting guidelines (these are core requirements for all organizations); (2) Sector supplements (which provide additional information from different sectors, if such information is available); (3) Technical protocols (these provide details of individual indicators, their definition, formulae and cross-referencing to minimize problems in comparability); (4) Issue guidance documents (which are non-sector-specific issues affecting a range of organizations, such as "diversity" and "productivity").

Väyrynen et al. (2014) described procedures of how customers of the manufacturing industry are now able to evaluate the entire supplying network working in one place, and it is equipped with equal requirements for quality and management. Moreover, they apply the similar methods to other key stakeholders—the regulating society, shareholders, employees, unions of employees and confederations of employers, in other words, social partners. Furthermore, the diverse driving forces have been present and active in developing HSEQ (health, safety, environment, quality) AP, i.e. the situation towards an equal level of quality and shared requirements of management for all companies operating at the sites and within a network of each business-leader manufacturing industry company (Väyrynen et al. 2014). HSEQ AP features key indicators describe both resulting outcomes and enablers of holistic excellence as far as a total of social, economic and ecological sustainability (Väyrynen et al. 2014).

Drucker (2001, p. 59) concluded that "the strength of business is accountability and measurability". Furthermore, Drucker (2001, p. 60) summarized this concept in the epigram "whoever claims authority thereby assumes responsibility; but whoever assumes responsibility thereby claims authority".

2 The Aim of the Study and Research Questions

The aim of the present study is to explore (1) the different kinds of practices implemented by Finnish corporations with respect to the "corporate sustainability" and how (2) the "corporate sustainability" is applied in their business practices. The corporate applications of the "corporate sustainability" will be investigated in the following areas: (1) Business strategy of the corporations, (2) Financiers and shareholders, (3) Customers and consumers, (4) Employees of the corporations and (5) Communities and authorities. A further aim is to describe and explain the cross-sectional association mechanisms, contexts and outcome patterns of the impact of the "corporate sustainability".

3 Research Materials and Research Methods

3.1 Research Materials

The data from this discourse study were produced between January and July 2014 from seven Finnish Corporations (see References, Corporation#1 2014, Corporation#2 2014, Corporation#3 2014, Corporation#4 2014, Corporation#5 2014, Corporation#6 2014, Corporation#7 2014).It originated from reviewing the online strategy reports on sustainability prepared by seven Finnish corporations; these were classified in this study into the taxonomies (based on theories of Stakeholder, Stewardship, Five Disciplines and Actantial Model) as follows: (1) business strategy of corporations, (2) financiers and shareholders, (3) customers and consumers, (4) employees of the corporations and (5) communities and authorities.

3.2 Research Methods

3.2.1 The Framework of the Theories Applied in Analysing Discourses

Discourse Analysis

In the present study, the sustainability reports of the Finnish corporations are explored as discourses. In general, discourse analysis can be defined as a set of methods and theories for investigating language in social contexts (Wetherell 1998; Wetherell et al. 2001). From this perspective, discourse is not "merely about actions, events and situations, it is also a potent and constitutive part of those actions, events and situations" (Potter 1997, 2000, p. 144). Discourse analysis is concerned with "understanding the nature of power and dominance" and how "discourse contributes to their production" (van Dijk 1995, 2001; Fairclough 1996, 2005). There are four broad sets of research issues which can productively be addressed specifically by discourse analysts in transdisciplinary research on organizational change, i.e. the problems of emergence, hegemony, recontextualization and operationalization (Fairclough 2005). In general, the discourse classification is based on Fairclough's (1996, p. 64) constructivism theoretical framework. Key assumptions can be summed up as follows: (1) discourses prepared by the corporation reflect a practical reality; (2) business experts can use a range of discourses that may be parallel and compete with each other; (3) the practical action is context bound; and (4) discourse is built on strategies and policies. Discourse is a dual concept. First, it represents the interactive process in which implications are produced, and secondly, it refers to the final outcome of this process (Fig. 1).

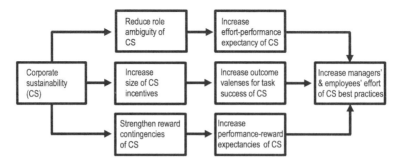

Fig. 1 The reference frames of the discourse analysis

Constructionism Approach

In the present study the viewpoints and text about the "corporate sustainability" published by Finnish corporations are assessed as the abilities to co-create the contexts which must be implemented by the corporation and/or by others. Similarly, they might shape other 'social realities' such as legitimacy, i.e. concepts that that often vary based on how the context is being constructed and through the discourse in achieving the goal. Constructionist researchers start with the assumption that access to shared dynamic and changing and individually constructed reality is only achieved through social constructions such as language and shared meanings (Eriksson and Kovalainen 2008).

The method of the data analysis of the empirical material described in the appendices is based on a constructionist approach. Consistent with the social construction of reality (Lord and Dinh 2012), these approaches emphasize the interactions themselves as well as the shared patterns of meaning-making, conjoint agency and coordinated behaviour through which the "corporate sustainability" is enacted. In this respect, the social interaction creates an understanding of collective efficacy and builds a different construct—an assessment of unique organization processes, rather than a simple aggregation of individual-level attributes (Cunliffe 2008). Thus, conscious experience is hypothesized to have a synergistic effect that can integrate constructs like current perceptions, goals, self-structures and experienced impacts to produce a coherent entirety (Fairhurst and Grant 2010). Constructionism assumes that social reality is not separate for different individuals, but is intimately interwoven and shaped by everyday interactions (Uhl-Bien et al. 2012).

The construction of innovations in the "corporate sustainability" and CSR can be understood as activities of translation, in which ideas are objectified into new forms and practices that fit the local institutional context (Windell 2006, p. 41).

Case Study

The present examined the Finnish corporations as cases to be explored. In this multiple-case study, the goal of the classification is to build a general explanation that suits each of the individual cases, even though the types of the "corporate sustainability" vary in their details. This kind of objective is analogous to conducting multiple experiments (Yin 1994, p. 112). The descriptions and analysis of multiple cases provide information on a more general phenomenon of the "corporate sustainability". Yin (1994, p. 13) defined that "A case study is an empirical inquiry that investigates a contemporary phenomenon within its real-life context, especially when the boundaries between phenomenon and context are not clearly evident". The case study research consists of a detailed investigation of organizations or groups within organizations, with a view to providing an analysis of the context and processes involved in the phenomenon under study with respect to the "corporate sustainability". This study is a descriptive case study of corporations and how they assess the "corporate sustainability". It consists of exploratory, descriptive or explanatory case studies (Yin 1994, p. 4). The approach used in these case studies is that of 'pattern-matching', whereby information supplied by the corporations is related to certain theoretical propositions with respect to the "corporate sustainability". This research design embodies a 'theory' of the topic being studied, and will present the collected data, the strategies for analysing the data, conclusions based on the data and theory development. For this reason, it is essential to provide an extensive background to the theory development prior to the assessment of the case study data.

Yin (1994, p. 30) argued that the level at which the generalization of the case study results will occur is that of appropriately developed theories and stated that the analysis of case study evidence is one of the least developed and most difficult aspects of case studies. This study deals with evidence about the "corporate sustainability", and it presents compelling analytic conclusions ruling out alternative interpretations. According to Yin (1994, p. 13) the case study (1) copes with the technically distinctive situation in which there will be many more variables of interest than data points, and since one result (2) will rely on multiple sources of evidence, with data needing to converge in a triangulating fashion, and as another result, (3) it benefits from the prior development of theoretical propositions to guide data collection and analysis.

In the present study the theoretical propositions are (1) Fifth Discipline of Senge (1994), (2) the stakeholder theory and (3) stewardship theory. For example, a case study of the "corporate sustainability" may reveal its variations in terms of both content and context, depending on the perspective of different stakeholders. The approach used to evaluate the cases of these corporations is the concept of

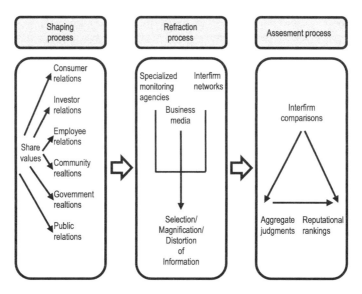

Fig. 2 Adapted from the concept of "reputation" as presented by Fombrun (2005) to the concept "corporate responsibility" within shaping practical process of CS, refraction process of CS and assessment process of the best practices of CS

"pattern-matching" described by Senge (1994), whereby several pieces of information about the "corporate sustainability" from the different corporation-related cases may be related to different practices (Fig. 2).

Classification of Discourses with Stakeholder Theory

Corporations pay a great deal of attention to the wishes and criticism of a widely divergent group of stakeholders about the "corporate sustainability". This requires a sound understanding of the impact of the values to which the various stakeholders adhere. The stakeholder theory and concept (Freeman 1984, 1994) are key elements in this study, as they outline the interdependence between society at large (stakeholders) and corporations—from a corporate perspective (looking outwards from inside the corporation). The stakeholder approach indicates that an organization is not only accountable to its shareholders but should also balance a multiplicity of stakeholders' interests that can affect or are affected by the achievement of the organization's objectives (van Marrewijk 2003). Freeman (1984) has stated that managers must satisfy a variety of constituents (e.g. workers, customers, suppliers, local community organizations) who can influence corporations' outcomes. According to this view, it is not sufficient for managers to focus exclusively on the needs of stockholders or the shareholder of the corporation. The stakeholder theory implies that it can be beneficial for the enterprise to engage in certain the "corporate sustainability" activities that non-financial stakeholders perceive as important, i.e. as without this activity, these groups might withdraw their support from the enterprise. The term 'stakeholder' can be defined both narrowly and broadly (Salzmann et al. 2006, p. 3). Salzmann et al. (2006) claimed that stakeholders' contributions to corporations include both tangible and intangible resources: shareholders provide capital; employees provide labour; customers bring loyalty and a certain willingness to pay; communities and governments are expected to establish a stable and beneficial regulatory framework, etc.

The stakeholder theory was expanded by Donaldson and Preston (1995) who stressed not only its moral and ethical dimensions but also the business case for engaging in these kinds of activities. According to Kujala's (2010) Finnish study, it is vital that corporations appreciate the importance of corporate responsibility issues in general, especially the meaning of stakeholder inclusion and enterprise–stakeholder interactions in business value creation, if they wish to achieve long-term corporate success. Stakeholders' authority can be based on several normative principles, such as public responsibility, legitimacy (Wood 1991) and sustainability (Wood and Jones 1995). Maon et al. (2010) argued that the stakeholder theory has emerged as crucial for understanding and describing the structures and dimensions of modern business and societal relationships. Thus, the theory helps to specify the groups or persons to whom corporations are responsible, and provides a foundation for legitimizing stakeholder influences on corporate decisions. Donaldson and Preston (1995) argued that the task of the stakeholder theory is to describe, and sometimes explain, specific corporate characteristics and behaviours. This theory can be considered as "managerial" since it recommends the attitudes, structures and practices which, when taken together, will constitute a stakeholder management philosophy (Donaldson and Preston 1995).

In line with the approach of Maon et al. (2010) on CSR, the present study builds on existing conceptualizations of the "corporate sustainability" according to the stakeholder and organizational culture-centred perspective. Thus, the present study describes the "corporate sustainability" according to Maon et al. characterizing CSR: (a) a stakeholder-orientated construct; (b) voluntary commitments of an organization; (c) issues extending inside and beyond the boundaries of that organization; and (d) driven by the organization's understanding and acknowledgement of its responsibilities regarding the impacts of its activities and processes on society. This approach also investigates whether there are key roles for organizational traits that will influence the corporate culture towards social responsiveness with regard to recognition of the "corporate sustainability" issues.

Since different stakeholder groups have different expectations and different information requirements about the "corporate sustainability", it is essential to investigate the practical ways that corporations can best implement its initiatives to promote the "corporate sustainability" with respect to their many target stakeholders. For example, Kujala (2010) in the Finnish study concluded that in order to enforce corporate responsibility, it should be possible to assure managers that there is a business case for stakeholder inclusion, and that stakeholder collaboration can create business models that are beneficial for all parties: the corporation and its owners, other stakeholders and society as a whole.

3.2.2 The Methods Applied in Analysing Discourses

Classification of Discourses with Five Disciplines by Senge

The developed relations and the categories that are treated as essential are verified over and over against the text and the data. One researcher moved continuously back and forth between inductive thinking (developing concepts, categories and relations from the text) and deductive thinking (testing the concepts, categories and relations against the text, especially against passages or cases that are different from those from which they were developed). The analyst must decide between equally salient phenomena and weigh them, so that one central category results together with the sub-categories, which are related to it (Flick 2006, p. 300). The core category again is developed in its features and dimensions and linked to other categories by using the parts and relations of the coding paradigm.

In the present approach the interpretation of discourse data cannot be regarded independent with respect to the collection of the material. Interpretation is the anchoring point for making decisions about which data or cases to integrate next in the analysis and how or with which methods they should be collected (Flick 2006, p. 296). In the process of interpretation, different 'procedures' for working with text can be differentiated. Texts serve three purposes in the process of qualitative research (Flick 2006, p. 83): not only are they the essential data on which findings are based, but also the basis of interpretations and the central medium for presenting and communicating findings. The coding provides different ways of handling textual material between which the researchers move back and forth if necessary and which they combine. Coding here is understood as representing the operations by which data are broken down, conceptualized and put back together in new ways.

The texts serve three purposes (Fig. 3): they represent (1) the essential data on which findings are based; (2) the basis of interpretations; and (3) the central medium for presenting and communicating findings (Flick 2006, p. 95).

The process of interpretation started with open coding, whereas towards the end of the whole analytical process, selective coding comes more to the fore. Starting from the data, the process of coding was leading to the detailed classification presented by Senge (1994) through a process. The coding strategies can be described that the researchers address the text regularly and repeatedly with basic questions (Flick 2006, p. 300):

(1) What? (What is the issue here? Which phenomenon is mentioned?)
(2) Who? (Which actors are involved? Which roles do they play? How do they interact?)
(3) How? (Which aspects of the phenomenon are mentioned?)
(4) When? How long? Where? Time, course, and location
(5) How much? How strong? Aspects of intensity
(6) Why? Which reasons are given or can be reconstructed?
(7) What for? With what intention, to which purpose? and
(8) By which? Means, tactics and strategies for reaching the goal.

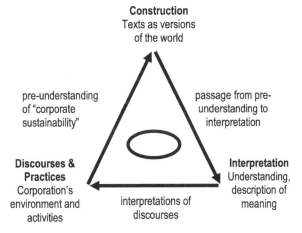

Fig. 3 Process of interpretation of "corporate sustainability" of this study (adapted from Flick 2006, p. 101)

In its ideal dimension, discourses of sustainability reports of the Finnish corporation cases perform both a cognitive function, by elaborating on the logic and contents of the reports, and a normative function, by demonstrating the appropriateness of the sustainability reports. In its interactive dimension, discourses perform a coordinative function by providing a common language and framework for the construction of the sustainability reports policy and a communicative function. The classification of the present study is based on the Five Disciplines of Senge (1994).

Senge (1994) presented a framework of the Five Disciplines of organizational learning as follows:

(1) Personal Mastery. This discipline of aspiration involves formulating a coherent picture of the results corporations that most sincerely wish to achieve, alongside a realistic assessment of their current situation. Learning to cultivate the tension between vision and reality can expand corporation's capacity to make better choices, and to achieve more of the goals that they have set for themselves.

(2) Mental Models. This discipline of reflection and inquiry skills focuses on developing awareness of the attitudes and perceptions that influence thought and interaction. By continually reflecting upon, talking about and reconsidering these internal perspectives of their environment, corporations can improve their ability to govern their actions and apply their making process more effective.

(3) Shared Vision. This collective discipline establishes a focus on mutual purpose. Individuals in corporations learn to nourish a sense of commitment to the group or organization by developing shared images of the future that they seek to create, and the principles and guiding practices which they hope will lead them to that future.

(4) Team Learning. This is a discipline of organizational interaction. There are different techniques such as dialogue and skilful discussion, which encourage teams to transform their collective thinking, and learn to mobilize their

energies and abilities in a synergistic manner. The overall outcome of the team is greater than the sum of their individual talents.

(5) Systems Thinking. In this discipline, individuals in organizations learn to better understand interdependency and change, and in that way they start to deal more effectively with the forces that shape the consequences of their actions. Systems thinking is based upon a growing theory regarding the behaviour of feedback and complexity—the innate tendencies of a system that lead to growth or stability over time.

Senge (1994, p. 359) stated that each of the five learning disciplines can be established at two distinct levels: practices (what you do) and principles (guiding ideas and insights). The practices are the activities upon which practitioners of the discipline focus their time and energy (Senge 1994, p. 359). Equally central to any discipline are the underlying principles, which represent the theory that lies behind the practices of the disciplines (Senge 1994, p. 374). Thus, mastering any of the disciplines requires effort at both levels—understanding the principles and following the practices. Learning always involves new understandings and behaviours; 'thinking' and 'doing' which is why principles are distinguished from practices (Senge 1994, p. 374).

The role of the general classification strategy presented by Senge (1994) is intended to help in choosing from different corporation-related practices and to complete the analytic phase of the evaluation of the "corporate sustainability". The role of the classification of the "corporate sustainability" is to determine the best ways of contrasting any differences as sharply as possible, and to develop theoretically significant explanations for the different outcomes. The present classification building represents a narrative form of the "corporate sustainability".

Classification of Discourses and Contents with Actantial Model of Greimas

In the data analysis of the discourses used by the Finnish corporations, the actantial model of Greimas (1987, 1990) was applied to provide a summary of the outcomes.

The six actants are divided into three oppositions, each of which forms an axis of the actantial description (Hébert 2011):

- The axis of goal: (1) subject/(2) object. The subject is the force directed towards an object. The relationship established between the subject and the object is called a junction.
- The axis of power: (3) helper/(4) opponent. The helper assists in achieving the desired junction between the subject and object; the opponent hinders this goal.
- The axis of transmission and knowledge: (5) sender/(6) receiver. The sender is the element requesting the establishment of the junction between subject and object. Sender elements are often receiver elements as well.

The core of the actantial model is formed from the subject's intention to acquire the object. Consequently, the subject is the performer of a task; the subject tries to acquire the object through some activity (Greimas and Courtes 1982). A sender is a party that gives the task to the subject (Greimas 1987, 1990). At the same time, the sender defines the value targets for the activity, and provides justification and motivation for the subject's activity. A receiver is a party who evaluates whether the subject has succeeded, and confers on the parties to the activities either a reward or punishment after the task has been completed.

4 Discourse Results of the Finnish Corporation Cases

According to their sustainability reports, it seems that most Finnish corporations (see References, Corporation#1 2014, Corporation#2 2014, Corporation#3 2014, Corporation#4 2014, Corporation#5 2014, Corporation#6 2014, Corporation#7 2014) are compliant with the following: GRI "Global Reporting Initiatives", ISO 9001 (2004) "Quality management systems", ISO 14001 (2008) "Environmental management systems" and BS OHSAS (2007) 18001 "Occupational health and safety management systems", UN "United Nations Global Compact", ILO (2014) Conventions and Recommendation and Decent Work Agenda, OECD "Guidelines for Multinational Enterprises", GRI (2014) "Global Reporting Initiatives" and AA1000APS (2008) "Account Ability Principles Standard".

Within framework of Five Disciplines (Senge 1994), the discourses utilized in the Finnish corporation cases as they described how they view the "corporate sustainability" have been subdivided into the following: (1) Business strategy of the corporations, (2) Financiers and shareholders, (3) Customers and consumers, (4) Employees of the corporations and (5) Communities and authorities. In the evaluations of the discourses supplied by the Finnish corporation cases, the actantial model of Greimas (1987, 1990) was applied to provide a summary of the discourses.

4.1 Discourses in Relation to the Business Strategy of the Corporations

4.1.1 Business Strategy of the Corporations: A Five Disciplines Classification

The major themes raised in the discourses of the Cases of the Finnish corporations how they would apply the "corporate sustainability" in relation to the 'Business strategy of the corporations: A Five Disciplines classification' are presented in Table A.1 (Appendix) with respect to seven Finnish Corporations (see References, Corporation#1 2014, Corporation#2 2014, Corporation#3 2014, Corporation#4 2014, Corporation#5 2014, Corporation#6 2014, Corporation#7 2014).

The examples of these discourses are presented as follows:

1. Personal Mastery
"We plan our environmental investments to increasingly apply the best available technologies for mitigating emissions."
"The renewed targets emphasise our role in society and measure not only environmental targets, but also the reputation, the customer satisfaction, and the security of supply chain."

2. Mental Models
"We believe that an ethical approach will lead to successful business, foster accountability and enhance our good reputation."
"A corporation that is financially strong is able to be responsible for the environment, take care of its personnel and meet the needs of its customers."

3. Shared Vision
"We have also further developed our processes for stakeholder engagement and gathering weak signals in order to [...] to assess what effects the global sustainability megatrends will have on our corporation."
"Sustainability is an integral part of the strategy. In its operations, the strategy gives balanced consideration to economic, social and environmental responsibility."

4. Team Learning
"Our focus is to drive continuous learning and leadership development to ensure professional growth."
"Mentoring is also used to transfer tacit knowledge from more experienced specialists to newcomers or less experienced employees."

5. Systems Thinking
"Our economic responsibility includes competitiveness, performance excellence and market-driven production, which create long-term value and enable profitable growth."
"Our corporate responsibility principles cover all aspects of our operations and strategy and are also integrated into the way that we conduct our business."

4.1.2 Business Strategy of the Corporations: An Actantial Model

Figure 4 presents the semiotic actantial model developed by Greimas (1987, 1990) which here is used to link the theory and the discourse data used in the business strategies of the corporations.

4.2 Discourses in Relation to Financiers and Shareholders

4.2.1 Financiers and Shareholders: A Five Disciplines Classification

The major themes raised in the discourses of the Cases of the Finnish corporations about how the "corporate sustainability" should be handled in relation to 'Financiers and shareholders: A Five Disciplines classification' are presented in Table A.2 (Appendix) with respect to seven Finnish Corporations (see References,

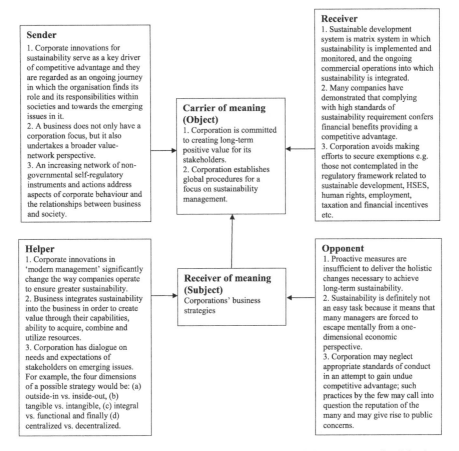

Fig. 4 Actantial model of the "corporate sustainability" in relation to corporations' business strategies

Corporation#1 2014, Corporation#2 2014, Corporation#3 2014, Corporation#4 2014, Corporation#5 2014, Corporation#6 2014, Corporation#7 2014).

The examples of these discourses are presented as follows:

1. Personal Mastery

"We have renewed our website, which is one of our main tools for sharing information with our stakeholders, including investors and financiers."

"Our multi-disciplinary team of corporate responsibility and assurance specialists possesses the requisite skills and experience within financial and non-financial assurance, corporate responsibility strategy and management […]"

2. Mental Models

"We have selected the indicators that are the most relevant to our operations, products and stakeholders based on the materiality analysis, an assessment of the most significant sustainability issues."

"We have particularly focused on developing the information content and presentation of our interim reports in order to provide information to investors."

3. Shared Vision
"We want to create value both for our shareholders and the society around us by finding solutions"
"We have carried out an online questionnaire about investor relations conduct for key owners, investors and analysts about their expectations […]"

4. Team Learning
"Our business divisions each additionally worked to establish their own websites providing stakeholders with division-specific information."
"Corporate Sustainability plays a central role in our market outlook and public affairs processes and supports investor relations."

5. Systems Thinking
"An Enterprise Risk Management (ERM) process provides a systematic approach for identifying threats and opportunities […]"
"Our aim of the new sustainability Key Performance Indicators (KPIs) is to secure the continual improvement and more frequent systematic monitoring of the progress."

4.2.2 Financiers and Shareholders: An Actantial Model

Figure 5 utilized the semiotic actantial model developed by Greimas (1987, 1990), connecting the theory and the discourse data provided by the corporation to financiers and shareholders.

4.3 Discourses in Relation to Customers and Consumers

4.3.1 Customers and Consumers: A Five Disciplines Classification

The major themes raised in the discourses of the Cases of the Finnish corporations about 'the "corporate sustainability"' in relation to 'Customers and consumers: A Five Disciplines classification' are presented in Table A.3 (Appendix) with respect to seven Finnish Corporations (see References, Corporation#1 2014, Corporation#2 2014, Corporation#3 2014, Corporation#4 2014, Corporation#5 2014, Corporation#6 2014, Corporation#7 2014).

The examples of these discourses are presented as follows:

1. Personal Mastery
"The aim is to ensure that the Supplier Code of Conduct is followed in practice throughout the supply chain."
"We are committed to operating in an ethically sound manner through responsible business practices across the entire value chain."

2. Mental Models
"Transparency and responsibility of the entire supply chain are of increasing interest to our customers and other stakeholders."
"We work to guarantee the best return on the customer's investment with minimal ecological impact."

14 A Case Study About the Discourses of the Finnish Corporations ...

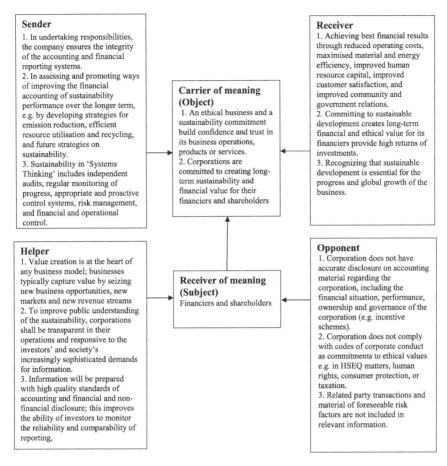

Fig. 5 Actantial model how the "corporate sustainability" applies in relation to the way that corporations communicate with 'Financiers and shareholders'

3. Shared Vision

"The requirement for transparency and responsibility of the supply chain comes also from our customers and other stakeholders."

"Continuous feedback and interaction with customers help us to improve our understanding of their needs, the challenges they face and the business environment our customers operate in."

4. Team Learning

"Together with our partners, we secure a sustainable supply of raw materials for our units and a supply of renewable products for our customers."

"We support our customers in reducing the ecological footprint of their operations."

5. Systems Thinking

"The Life Cycle Assessment (LCA) assesses the environmental impacts associated with all stages of a product's life cycle."

"Integrated management system covers quality, safety, energy and environmental aspects."

4.3.2 Customers and Consumers: An Actantial Model

Figure 6 presents the semiotic actantial model developed by Greimas (1987, 1990), connecting the theory and the discourse data used in their relationships with customers and consumers.

4.4 Discourses in Relation to Employees of the Corporations

4.4.1 Employees of the Corporations: A Five Disciplines Classification

Major themes raised in the discourses of the Cases of the Finnish corporations about how the "corporate sustainability" is handled in relation to 'Employees of the corporations: A Five Disciplines classification' are presented in Table A.4

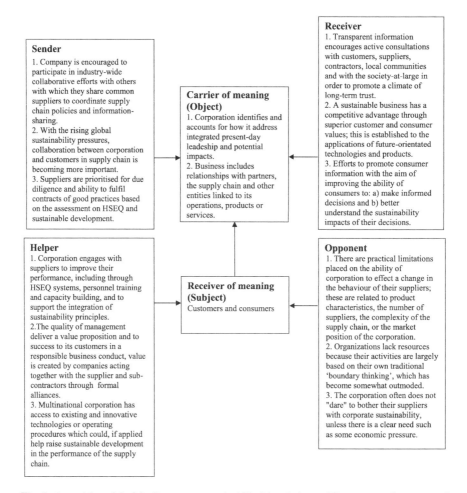

Fig. 6 Actantial model of the "corporate sustainability" in relation to 'Customers and consumers'

(Appendix) with respect to seven Finnish Corporations (see References, Corporation#1 2014, Corporation#2 2014, Corporation#3 2014, Corporation#4 2014, Corporation#5 2014, Corporation#6 2014, Corporation#7 2014).

The examples of these discourses are presented as follows:

1. Personal Mastery
"We offer our employees the possibility to enhance their competences through training, internal job rotation and other learning opportunities."
"We encourage our OHS networks to share best practices, learn from any incidents, clarify modes of operation, roles and responsibilities, and improve reporting practices."

2. Mental Models
"We promote a risk-aware culture in all areas of the company's decision-making."
"Improving the sustainability of our own operations, supply chain management, and technology and plant safety management have been identified as important targets to meet the expectations of customers, employees and other stakeholders."

3. Shared Vision
"Safety reflects the overall quality of our operations. Safety training is systematic and consistent […]"
"Investing in our people is our strategic priority, whether this involves ethical leadership, safety issues […]"

4. Team Learning
"We constantly educate our employees and management to create a healthy working environment […]"
"A new IT-tool is implemented to help us to identify, categorise, assess and monitor our suppliers […]"

5. Systems Thinking
"We apply our safety toolbox in all units. This toolbox includes practical, hands-on tools that directly address individuals' behaviour."
"Sustainable management is based on our policies and commitments as well as instructions […]"

4.4.2 Employees of the Corporations: An Actantial Model

Figure 7 presents the semiotic actantial model developed by Greimas (1987, 1990), linking the theory and the discourse data about the "corporate sustainability" in the way that the corporations relate to their employees.

4.5 Discourses in Relation to Communities and Authorities

4.5.1 Communities and Authorities: A Five Disciplines Classification

Major themes raised in the discourses of how the Finnish corporations consider the "corporate sustainability" in relation to 'Communities and authorities: A Five Disciplines classification' are presented in Table A.5 (Appendix) with respect to

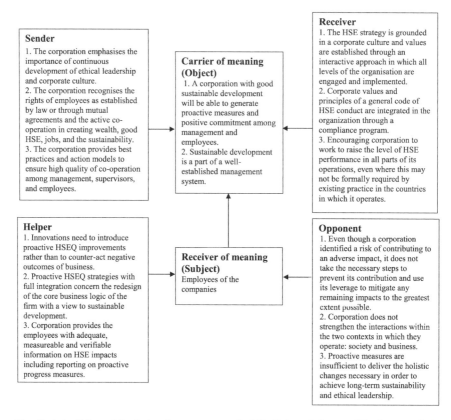

Fig. 7 Actantial model of the "corporate sustainability" in relation to 'Employees of the corporations'

seven Finnish Corporations (see References, Corporation#1 2014, Corporation#2 2014, Corporation#3 2014, Corporation#4 2014, Corporation#5 2014, Corporation#6 2014, Corporation#7 2014).

The examples of these discourses are presented as follows:

1. Personal Mastery
"We generate well-being at work, in local communities and throughout society at large, and are committed to global sustainability principles."
"We implement open and timely communication with the local community."

2. Mental Models
"We recognize the need to create shared value in the communities where we operate."
"Our investments develop e.g. sustainable production capacity, energy efficiency, local infrastructure and electricity distribution reliability."

3. Shared Vision
"One of our most fundamental duties is to guarantee that all our operations are sustainable and all our products are safe for both people and the environment."
"Our aim is to be an open and equal work community."

4. Team Learning
"Our social responsibility includes being a good corporate citizen [...]"
"We obtain stakeholder feedback largely through our regular contacts with our stakeholders at [...] public hearings, open house events [...]"

5. Systems Thinking
"We aim at completing community projects jointly with our customers with joint financing."
"The aim is long-term partnership with the community."

4.5.2 Communities and Authorities: An Actantial Model

Figure 8 presents the semiotic actantial model developed by Greimas (1987, 1990), connecting the theory and the discourse data of the "corporate sustainability" the way that corporations relate to neighbouring "Communities and authorities".

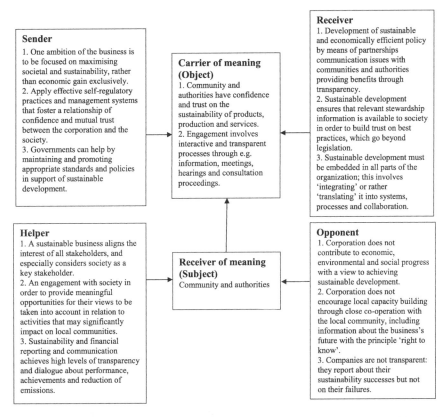

Fig. 8 Actantial model of the "corporate sustainability" in relation to "Communities and authorities"

5 Discussion

This study supports the results of the other evaluators (e.g. Christmann 2000; Wagner 2005) when they claim that the proactive corporate strategies that go beyond regulatory compliance have a positive effect on corporate performance. In all business sectors, stakeholder pressure has often been cited as a factor contributing to the adoption of proactive environmental practices by commercial enterprises (Wheeler et al. 2003). The present results support the findings of other investigators (e.g. Webley and Werner 2008) who indicated that a sustaining corporate culture should include continuous training, employee engagement and consistent communication. In addition, by allowing employees voice in organizational decisions, using rewards to encourage ethical behavior and injecting ethical values in regular business activity, ethical leaders enrich the autonomy and significance of work (Piccolo 2010).

This study supports the conclusion of Zink (2014) that there is a need for systems thinking for 'sustainability' to take into consideration, e.g. the following: (1) the human and social capital, e.g. skills, knowledge, safety, health, motivation, participation, trustworthiness; (2) the frame of changes in, e.g. demographics, globalization; (3) a systemic and holistic approach regarding whole value creation and supply chains; (4) a life-cycle-perspective regarding products and work systems (design process, manufacturing, assembly, maintenance and repair, disassembly, reuse and recycling); and (5) the inclusion of organizational systems, e.g. leadership and corporate culture.

This study on the "corporate sustainability" is also in line with the results of Du et al. (2010) on CSR. The present results highlight that the fact the surveyed corporations are committed to the "corporate sustainability" since it creates long-term value for their customers, shareholders, employees and society. Furthermore, the present corporations' aims are to meet their stakeholders' needs effectively. This study revealed that these corporations also cooperate with governmental authorities in the effective development and implementation of regulations and standards, which they not only meet but often go beyond. Du et al. (2010) conclude that by investing in social initiatives, a corporation will be able not only to generate favourable stakeholder attitudes and behaviours, but also, in the long run, to build a corporate/brand image and strengthen stakeholder–corporation relationships. According to a Finnish study on CSR (Kujala 2010), rather than emphasizing short-term profit making, managers appear, at least to some extent, to understand the importance that long-term success for the corporation has to be achieved by balancing economic interests and stakeholder issues.

This study on the "corporate sustainability" supports the results of Maon et al. (2010). There are certain requirements which need to be taken into consideration when trying to integrate sustainable principles into business models and processes. These best practices generally rely on the idea of a level-by-level process along with internal capabilities gradually applied to strategy issues. The present findings show that one can devise a set of performance indicators related to the "corporate sustainability" with which improvements can be measured, and the company can introduce and apply systematic procedures to verify the implementation of measurable elements of the "corporate sustainability". During the last development stage of the "corporate sustainability", the corporation will have fully integrated the principles of the "corporate sustainability" into every aspect of the organization and its activities.

Further research and development innovations for the "corporate sustainability" are needed in order to utilize HSEQ assessment procedures. Väyrynen et al. (2012) have developed and started to apply HSEQ Assessment Procedure (HSEQ AP) for measuring and evaluating their suppliers within large-scale process and other manufacturing industry companies in Finland. The general principles and practices behind HSEQ AP, in addition to the ones developed by local research and development, has comprised European Union and national legislation, Total Quality Management (TQM), Excellence and Quality Award Models, social responsibility, and a total of social, economic and ecological sustainability (Väyrynen et al. 2014).

5.1 Theoretical Implications

Weick (2002, p. 11) found that organizational action is based on interpretations of discourses; these interpretations are externalized by concrete activities. Learning to discern underlying 'structures' rather than 'events' is a starting point; each of the "systems archetypes" developed points to areas of high- and low-leverage change (Senge 1994, p. 65). Schein (1996, p. 64) has concluded that the culture of the "corporate sustainability" has 'changed'—in reality, it has become enlarged—through changes in various key concepts in the mental models of individuals who are the main carriers of the culture. Furthermore, the essence of the discipline of systems thinking lies in a shift of mind: discerning inter-relationships rather than linear cause–effect chains, and seeing processes of change (Senge 1994, p. 73).

The present findings on the stakeholder theory support the views of Salzmann et al. (2006, p. 3) who claimed that corporations exist in permanent contact with their economic, legal, political, societal and technological environments. The present results on stakeholder theory are consistent with the approach of Jones (1995), who stated that the corporations view their repeated transactions with stakeholders as ways to emphasize the basis of trust and cooperation, i.e. incentives to be honest and ethical, since such behaviour is beneficial to the company.

Donaldson and Preston (1995) argued that stakeholder theory is used to identify the connections, or lack of connections, between stakeholder management and the achievement of traditional corporate objectives (e.g. profitability, growth). At present, the studied corporations are committed to creating long-term value for their shareholders. Sustainable financial performance, responsible behaviour as a corporate citizen and ongoing improvement in HES areas are the practical dimensions of present the "corporate sustainability" together with CSR.

The present results support the conclusions of Senge (1994, p. 73) that the practice of systems thinking starts with understanding a simple concept called 'feedback' that demonstrates how actions can reinforce or counteract (balance) each other. In this context, the feedback perspective suggests that everyone shares responsibility for problems generated by a system: this does not necessarily imply that everyone involved can exert equal leverage in changing the system (Senge 1994, p. 78).

This study showed also that many of the practices most conducive to developing one's own personal mastery are embedded in the disciplines for building learning organizations (applied from Senge 1994, p. 173). Furthermore, the most positive actions that an organization can take to foster personal mastery involve working to develop all five learning disciplines in the "corporate sustainability". In addition, the tools of systems thinking are also important because virtually all the prime tasks of management teams are to develop strategies, shape visions, to devise policies and to design organizational structures (Senge 1994, p. 256). This study found that in a learning organization, leaders are responsible for building organizations where both the managers and the employees continually expand their capabilities to simplify complexity, to clarify vision, to improve shared mental models and to implement best practices in the "corporate sustainability", i.e. they are responsible for continuous learning processes (applied from Senge 1994, p. 340). In a large organization, different combinations of learning disciplines will be developing in different operating units (Senge 1994, p. 344). In addition, leadership has to operate at many levels, from local leaders to global leaders but the leadership should strive to introduce the disciplines implementing proactive solutions to current problems in global issues with respect organization-wide learning processes (applied from Senge 1994, p. 344).

5.2 Practical Implication

The present results support the conclusions of Drucker (2001, p. 11) that every enterprise requires commitment to common goals and shared values. The enterprise must have simple, clear and unifying objectives. Furthermore, the mission of the organization has to be sufficiently clear enough and large enough to provide common vision. Furthermore, Drucker (2001, p. 29) showed that objectives have to be set in these eight key areas: (1) Marketing, (2) Innovation, (3) Human resources, (4) Financial resources, (5) Physical resources, (6) Productivity, (7) Social responsibility and (8) Profit requirements.

This study supports the conclusions of Jackson (2003, p. 25) that the "corporate sustainability" in an interactive planning seeks to gain stakeholder approval for and commitment to an 'idealized design' for the system being developed. In addition, this is meant to ensure that the maximum creativity is brought to the process of the "corporate sustainability" by replacing the current messy situation with which stakeholders are confronted and replacing it with a clear future that they all desire.

Appropriate means for achieving the idealized design of the "corporate sustainability" are then sought. Soft systems methodology enables managers to work with and change the "corporate sustainability" value systems and cultures that exist in organizations (applied from Jackson 2003, p. 25). Furthermore, Christensen (1997, p. 175) showed that when new challenges require different people or groups to interact differently than they habitually have done—addressing different challenges with different timing than historically had been required—managers need to pull the relevant people out of the existing organization and draw a new boundary around a new group. In this context new team boundaries enable or facilitate new patterns of working together that ultimately can coalesce as new processes—new capabilities for transforming inputs into outputs (Christensen 1997, p. 175). In these "heavy-weigh-teams" members do not simply represent the interests and skills of their function. They are charged to reach decisions and make trade-offs for the good of the project (Christensen 1997, p. 177).

The present results support the findings of Senge (1994) who stated that in system dynamics the most important tool that organizations must master is the route to becoming 'learning organizations'. Only system dynamics can reveal the systemic structures that govern their behaviour. The basic purpose of leadership is to extend the practice of the three core values, 'Merit, Openness and Localness', by highlighting the skills needed to put them into practice (Senge 1994, p. 183). For example, building shared sense establishes the most basic level of commonality in visions and operating values of an organization (Senge 1994, p. 208). If openness is a quality of relationships, then building relationships characterized by openness may be one of the most high-leverage actions to help in building the relationships of organizations (Senge 1994, p. 208).

The present results support the conclusions of Yukl (1998, p. 61) that the best way to group specific business behaviours into general categories would be as follows: (1) Task-oriented behaviour; doing things that are primarily concerned with accomplishing the task, utilizing personnel and resources efficiently, maintaining stable and reliable operations, and making incremental improvements in quality and productivity; (2) Relationship-oriented behaviour; doing things that are primarily concerned with improving relationships, increasing cooperation, increasing job satisfaction of subordinates, and building engagement with the organization; (3) Change-oriented behaviour; doing things that are primarily concerned with improving strategic decisions, adapting to changes, making major changes in objectives, processes, or products/services, and gaining commitment to the changes. Furthermore, as applied from Yukl (1998, pp. 266–267) the "corporate sustainability" should include the following aspects: (1) Supportive leadership, (2) Directive leadership with specific guidance, procedures, scheduling and coordinating the work, (3) Participative leadership and (4) Achievement-oriented leadership with challenging goals, improvements and performance excellence.

This study supports the conclusions of Schein (1996, p. 68) that leaders of the future will have to possess more of the following characteristics: (1) Extraordinary levels of perception and insight into the realities of the world and into themselves, (2) New skills in analysing cultural assumptions, identifying functional assumptions, and evolving processes that enlarge the culture by building on its strengths and functional elements, (3) The willingness and ability to involve others and elicit their participation and (4) The willingness and ability to share power and control according to people's knowledge and skills. Work (1996, p. 78) described his concept of true leadership, true leaders for the future must be willing to accept four fundamental challenges: (1) They must have a vision for the workplace that ultimately results in a significant broadening of the corporate culture and the workplace environment, (2) They must be willing to craft and implement new and different communication processes to enhance and promote perceptions of fairness and equity, (3) They must be willing to bring full and unquestioned commitment to the effective utilization of a diverse work force and (4) They must act as the linchpin between their organization and the larger community.

This study also supports the results of van Marrewijk (2003) that a higher ambition level and specific interventions with respect to the "corporate sustainability" require a supporting organizational framework and a proper value system (van Marrewijk 2003). Furthermore, when challenged by changing circumstances and confronted by new opportunities, individuals, organizations and societies have to develop adequate solutions that preferably create synergy and add value at a higher level of complexity (van Marrewijk 2003). In addition, those leaders who exhibit balanced processing solicit views from their followers, are indicating that they are willing to have their positions or beliefs challenged before coming to a decision (Hannah et al. 2011). By soliciting views from the stakeholder, leaders can promulgate better understanding with respect to abstract principles and ethical standards, and engage their followers in ethical issues.

Leeman (2002) highlighted the positive results of an interactive planning project conducted in the DuPont who adopted a commitment in their EHS policy. From this commitment, the corporation derived a new slogan, 'The Goal is Zero'. Furthermore, the openness of individuals and organizations to acknowledge an error or problem, and ensuring that learning ensues from the any incident, were found to more likely to reflect the depth of commitment to ethical practice than pious claims of having high ethical standards proclaimed in mission statements or buried in corporate reports (Fisher and Lovell 2006, p. 403).

5.3 Generalization of the Present Results

Unquestionably, there are some limitations concerning our respondents which must be noted and these limit the possibility to generalize the results to other Western Europe countries. According to Yin (1994, p. 34), if one wishes to evaluate of construct validity (establishing correct operational measures for the concepts being studied), an investigator must consider the following two steps: (a) select the specific types of changes that are to be studied; and (b) demonstrate that the selected measures of these changes do indeed reflect the specific types of change that have been selected. There are three tactics available to increase construct validity

(Yin 1994, p. 34). The first is the use of multiple sources of evidence. The second tactic is to establish a chain of evidence. The third way is to have the draft case study report reviewed by key informants (Yin 1994, pp. 34–35).

The particular strength of case studies lies in their capacity to explore social processes as they unfold in organizations. However, the key feature of the case study approach is not the method or the data it generates but the emphasis on understanding processes as they occur in their context. The emphasis is not on divorcing context from the topic under investigation but rather on viewing this as a strength. This study tries to make explicit arguments and it allows the readers to consider the generalization claim. It is the reader who on the basis of the detailed contextual descriptions of the qualitative results, has to judge whether the findings can be generalized to a novel situation (Kvale and Brinkmann 2009). Grouping the answers of the different stakeholders into a discourse analysis under the same headings increases the transparency of the present study i.e. the reader can compare and follow the contexts and dynamics in the responses of the verbal discourse repertoires used by the different stakeholders.

The findings of the qualitative study are judged to be reasonably reliable and valid, but the question remains whether the results are primarily of local interest, or whether they may be transferable to other subjects and situations (Kvale and Brinkmann 2009). The generalizability of the results refers to the question of whether the findings of this investigation are applicable to other situations. In qualitative studies, this question has to be approached differently because of the different assumptions about the nature of reality (Lincoln and Guba 1985). The term transferability is appropriate if one wishes to determine whether the findings have a wider significance (Guba and Lincoln 1989). Therefore, the way to provide transferability is to use reader generalizability, meaning that the outside reader has to assess whether the results can be applied to his or her own context. In order for that to happen, the researcher must provide enough information for the reader to estimate whether the findings can be transferred to his or her own settings.

The concept of validity has been described in qualitative studies by the use of a wide range of terms. Validity is not a single, fixed concept, but rather a contingent construct, inescapably based on the processes and intentions of particular research methodologies and projects (Eriksson and Kovalainen 2008). An aggregated definition of 'validity' could be accuracy, and the definition of 'reliability' could be replicability (Denzin and Lincoln 1998). The criteria for judging the quality of research designs have been summarized as follows (Yin 1994): (1) Construct validity: establishing correct operational measures for the concepts being studied, (2) Internal validity: establishing a causal relationship, whereby certain conditions are shown to lead to other conditions, as distinguished from spurious relationships, (3) External validity: establishing the domain to which a study's findings can be generalized, and (4) Reliability: demonstrating that the operations of a study—such as the data collection procedures, can be repeated and would achieve the same results.

With respect to reliability, validity and replicability discursive research can utilize four different criteria: fruitfulness, quality of interpretation, quality of transcription and usefulness (Kvale and Brinkmann 2009). This study makes the arguments explicit and allows the readers to consider the generalization claim. The reader on the basis of the detailed contextual descriptions of the qualitative results has to judge for him/herself whether the findings may be generalized to novel situations (Eriksson and Kovalainen 2008). In my view, grouping the answers of the different responses in the discourse analysis under the same headings increases the transparency i.e. the reader can compare and follow the contexts and dynamics in the responses of the textual discourses.

Discourses also construct objects (Lord and Dinh 2012; Cunliffe 2008; Fairhurst and Grant 2010) i.e. the respondents create their own organizational realities by adoption of specific discourses. Discursive scholars (Lincoln and Guba 1985) represent a constellation of perspectives united by the view that language does not mirror reality, but constitutes it. Human communication is also more than a simple act of transmission; it is concerned with the construction and negotiation of meaning (Uhl-Bien et al. 2012).

The limitations related to the applicability of Greimas' model are twofold. First, as an analytical tool, although Greimas' model may appear clear and simple, it may prove to be rather complicated as there are no definitions to identify the actants. The construction of the relationships between the actantial units of the text is always relative, and depends on the researchers' definitions. The semiotic analysis requires close assessment and a clear interpretation of the revealed meanings. However, despite the present limitations, the present results support the proposal that Greimas' model can be used as a theoretical framework and as practical semiotic tool describing the "corporate sustainability".

6 Conclusions and Recommendations

The values of an organization are the criteria by which decisions about priorities are made. Some corporate values are ethical in their nature, such as those that guide the decisions and especially the processes. Within the framework of "corporate sustainability", values have a broader meaning. An organization's values are the standards by which managers and employees make prioritization decisions. Successful companies' managers and employees gradually come to become that the "corporate sustainability" priorities of doing things and methods of making decisions are the right way to work. It seems that there is a need to build broad coalitions within the corporation that can be mobilized to address the best practices of the "corporate sustainability" issues. For example, systematic leadership towards "corporate sustainability" could imply utilization of systems thinking in order to take step-wise approaches to transformative changes. The quality of the management is the key because this will determine the success of the business model; it determines their capabilities, ability to acquire, combine and utilize valuable resources in ways that will deliver value to the stakeholder. Although all business sustainability innovations that deliver sustainability are welcomed, proactive innovation strategies appear to exert the greatest impact.

Appendix

Major themes raised in the discourses of the Cases of the Finnish corporations

Table A.1 Major themes raised in the discourses of the cases of the Finnish corporations about how they approach the "corporate sustainability" in relation to the 'Business strategy of the corporations: A five disciplines classification'

Corporation#1
1. Personal Mastery. "Our business areas follow high moral and integrity in advocacy activities as described in the Code of Conduct and Sustainability Principles. We implement actions that improve profitability and result in targeted growth, while enhancing resource efficiency and environmental performance"
2. Mental Models. "High environmental standards and safety performance are integral elements of our management system and are continuously improved"
3. Shared Vision. "We have further developed our processes for stakeholder engagement and gathering weak signals"
4. Team Learning. "We have developed a risk assessment process for our raw material suppliers in order to identify risk-rated suppliers that are to be audited from a sustainability point of view"
5. Systems Thinking. "The weak signals are gathered around these 12 global macro trends: climate change, energy, buildings, transport, waste, water, biodiversity, integrating sustainability, communication, ethical business practices, security and equality. We continue to invest significantly into making our processes more material efficient"
Corporation#2
1. Personal Mastery. "We plan our environmental investments to increasingly apply the best available technologies"
2. Mental Models. "We believe that an ethical approach will lead to successful business, foster accountability and enhance our good reputation"
3. Shared Vision. "Ethical business conduct and legal compliance are not only cornerstones of our way of doing business, but also paramount for living up to our corporation's purpose 'Do Good for the People and the Planet' and our motto 'Do What's Right'"
4. Team Learning. "We measure the sustainability performance of our operations and products using environmental, social and economic performance indicators and related targets. We set high ethical and professional standards"
5. Systems Thinking. "Our Key Performance Indicators (KPIs) are reviewed as part of our Global Responsibility Strategy. Our KPIs are structured by our Global Responsibility Lead Areas."
Corporation#3
1. Personal Mastery. "The renewed targets emphasise our role in society and measure not only environmental targets, but also the reputation, the customer satisfaction, and the security of supply chain. Our business operations and purchasing principles consist of the following:

(continued)

Table A.1 (continued)

(1) Compliance with the Code of Conduct, regulations and agreements, (2) Professional purchasing process, (3) Supplier pre-selection and audits; and (4) Monitoring of reputation development"
2. Mental Models. "Our economic responsibility emphasizes strong financial performance, profitable growth and added value over the long term. A corporation that is financially strong is able to be responsible for the environment, take care of its personnel and meet the needs of its customers"
3. Shared Vision. "Sustainability is an integral part of the strategy. In its operations, the strategy gives balanced consideration to economic, social and environmental responsibility"
4. Team Learning. "We support the development of society and produce added value for its different stakeholders. Strong financial performance and growth must be achieved in compliance with sustainability principles and the corporation's target setting"
5. Systems Thinking. "Our economic responsibility includes competitiveness, performance excellence and market-driven production, which create long-term value and enable profitable growth. Sustainable and environmental incidents and indicators measuring the reliability of power distribution are reported monthly. Sustainability targets are based on the continuous improvement of operations"
Corporation#4
1. Personal Mastery. "We prioritize management training and developing our strategic capabilities. We defined a new way of working model for personnel across known as 'The Way Forward'. Based on the corporation's values, this new initiative is intended to secure our ability to succeed in a changing world, both today and into the future, and leverage everyone's resources for the corporation's benefit"
2. Mental Models. "Developing coaching-oriented leadership will also help translate the thinking behind the new 'The Way Forward' programmes' initiative into practical measures and practical progress. Risk management framework is based on three risk assessment elements: (1) an Enterprise Risk Management process; (2) risk manuals for specific risk disciplines and (3) risk awareness across the organization, based on proactive thinking and behaviour among individual employees"
3. Shared Vision. "Our programme 'The Way Forward' is central to how we are working to achieve our strategic targets and make our corporation a more profitable, a more customer-focused, and a safer corporation and one where personnel enjoy their jobs and feel good about what they do"
4. Team Learning. "We believe that a safe workplace, challenging jobs, good management, and a business culture that encourages people to perform at their best are keys to the wellbeing of its personnel and the success of its business"
5. Systems Thinking. "We have brought our main strategic projects under the umbrella of four Value Creation programs: Profitable Growth, Productivity, Renewable Feedstock, and customer Focus"
Corporation#5
1. Personal Mastery. "Our goal with talent management and professional growth is to enable our employees to reach their full potential. Based on our materiality assessment, we have defined the areas where we want to reach the status of a forerunner or where we want to meet our stakeholders' anticipation level"
2. Mental Models. "Our sustainability agenda comprises four areas: our offering, our people, our community engagement and our governance"

(continued)

Table A.1 (continued)

3. Shared Vision. "Health, safety and social responsibility for employees, customers and the local community are becoming increasingly important. We have a clear hierarchy of ethics guidance and decision-making in sustainability issues"

4. Team Learning. "Our focus is to drive continuous learning and leadership development to ensure professional growth. Mentoring is used to transfer tacit knowledge from more experienced specialists to newcomers or less experienced employees"

5. Systems Thinking. "Six strategic programs were introduced in 2013. The first focuses on customer experience and aims to improve customer-centricity and delivery excellence. The second strengthens our earnings logic to capture more value with our technologies over the life cycle. The third program was made to ensure the technology leadership, and the fourth to increase cost competiveness of our products. The fifth program focuses on further developing common business processes and tools, and the sixth program is dedicated to our people and professional growth"

Corporation#6

1. Personal Mastery. "We continue to build on our strong heritage in sustainability, high product quality and solid technical expertise"

2. Mental Models. "We are more committed than ever to operating in an ethically sound manner through responsible business practices across the entire value chain. Performance and development dialogues (PDDs) support our strategy execution and reinforce a performance-driven mindset"

3. Shared Vision. "We continue to pursue transparency, learning from dialogue with customers, investors, non-governmental organizations and local communities"

4. Team Learning. "The essence of the process is engaging and involving our personnel in the strategy implementation. The business targets are cascaded into individual targets to ensure that actions are aligned with our business strategy and support our success"

5. Systems Thinking. "Our reporting reflects the view that all of our operations—and our dialogue with stakeholders—must be based on ethical and sustainable business practices, since these provide the basis for our long-term competitiveness"

Corporation#7

1. Personal Mastery. "Responsible business within the economic, social and environmental aspects is a requirement for sustainability competitiveness. Profitable business is built on responsible operations"

2. Mental Models. "We pursue growth from specialization and the emerging markets"

3. Shared Vision. "Our customers benefit from well-managed supply chain management, which consists of responsible sourcing, dependable logistics and the continuous development of our own operating activities"

4. Team Learning. "Expert people and well-managed human resources, environmental and safety affairs create the foundation for profitability. By focusing on energy and material efficiency, and by investing in environmental technology, we have successfully significantly reduced emissions over the past decades"

5. Systems Thinking. "Responsible operations are a key element of the strategic focus areas. Focus areas of corporate responsibility are as follows: (1) Supporting customer business by providing energy efficient solutions that reduce lifecycle costs, (2) Improved energy efficiency in own operations, (3) Increased product development and manufacturing technology competence"

Table A.2 Major themes by the Finnish case corporations about the "corporate sustainability" in relation to 'Financiers and shareholders: A five disciplines classification'

Corporation#1
1. Personal Mastery. "We continuously develop the productivity and efficiency of our production units with investments and development actions that improve profitability and result in targeted growth, while enhancing resource efficiency and environmental performance"
2. Mental Models. "We have selected the indicators that are the most relevant to our operations, products and stakeholders based on the materiality analysis, an assessment of the most significant sustainability issues"
3. Shared Vision. "We have developed further our processes on stakeholder engagement and gathering weak signals on megatrends potentially having impact on the corporation"
4. Team Learning. "We develop actions that improve profitability and result in targeted growth, while enhancing resource efficiency and environmental performance"
5. Systems Thinking. "The sustainability Report 2013 has been prepared according to the Global Reporting Initiative (GRI) guidelines"
Corporation#2
1. Personal Mastery. "We have renewed our website, which is one of our main tools for sharing information with our stakeholders, including investors and financiers"
2. Mental Models. "We have transformed ourselves into a value-driven corporation focused on growth markets. This process involves looking beyond the traditional ways of utilising raw materials and serving customers in new markets"
3. Shared Vision. "At the same time governments, shareholders, investors and NGOs have all started to put more emphasis on the need for corporations to holistically tackle human rights dilemmas by taking initiatives that go beyond meeting legal obligations"
4. Team Learning. "Our business divisions each additionally worked to establish their own websites providing stakeholders with division-specific information"
5. Systems Thinking. "Conducted prior to the investment decision, these assessments cover all relevant factors related to environmental, social and business practice issues"
Corporation#3
1. Personal Mastery. "Our corporation is financially strong and is able to shoulder its responsibility for the environment, take care of its personnel, meet the needs of its customers and support the development of the entire society"
2. Mental Models. "Our investments pursue a financially profitable balance that provides the possibility to increase capacity and reduce emissions"
3. Shared Vision. "Our Finance Forum deals with investments and gives sustainability approval (environmental, occupational health, sustainable and social impacts) for all significant investments, acquisitions and divestments as part of investment evaluation and approval procedure"
4. Team Learning. "Corporate Sustainability plays a central role in our market outlook and public affairs processes and supports investor relations. Our information to investors consists of (1) Expertise investments, (2) Maintenance, productivity and legislation-based investments, (3) Growth investments, (4) Research and development expenditure"
5. Systems Thinking. "Our investors' and shareholders' expectations in our financial position and ability to take care of the agreed obligations are the following: (1) High-yield share, (2) Risk management, (3) Responsible operations, (4) Dedicated to achievement of our financial targets, (5) Dividends paid every year, (6) Development of suppliers' business/products/services and (7) Good reputation"

(continued)

Table A.2 (continued)

Corporation#4
1. Personal Mastery. "Our multi-disciplinary team of corporate responsibility and assurance specialists possesses the requisite skills and experience within financial and non-financial assurance, corporate responsibility strategy and management, social and environmental issues, as well as knowledge of the energy industry, to undertake this assurance engagement"
2. Mental Models. "Our values and management systems are the foundation of the control environment and provide the background for shaping people's awareness and understanding of control issues"
3. Shared Vision. "With respect to financial reporting, our control measures cover: (1) the President, Chief Executive Officer (CEO) and corporate management and (2) the Audit Committee"
4. Team Learning. "The central task of Internal Audit is to audit the operations of our units and functions on a regular basis and evaluate their internal controls, risk management, and administrative practices. The areas to be audited are determined by the projected financial and operational risks concerned"
5. Systems Thinking. "An Enterprise Risk Management (ERM) process provides a systematic approach for identifying threats and opportunities related to strategic targets and business plans. The control measures cover the responsibilities for overseeing the financial reporting process and related controls as follows: (1) clearly defined financial reporting roles, responsibilities and (2) the resources allocated within financial reporting e.g. segregation of duties, competencies recruited and retained"
Corporation#5
1. Personal Mastery. "Our long-term target is all about our handprint: we must be able to offer more sustainable technologies and services to our customers with less harmful impact on the environment"
2. Mental Models. "Commitment to sustainability is our core value. It means that sustainability not only guides what we do, but also how we do it. The impact of our own operations on the environment is tiny compared to what we can achieve by providing innovative solutions to our customers; this is also how we measure our success"
3. Shared Vision. "We want to create value both for our shareholders and the society around us by finding solutions to some of the challenges our modern lifestyle pose for the planet and the generations to come"
4. Team Learning. "Our investment team has a continuous dialogue with investors and analysts and meets them on a regular basis at quarterly reporting investor meetings and capital market days. In order to develop our investor relations capabilities further, we conduct annual surveys amongst analysts and investors about their expectations"
5. Systems Thinking. "We integrate sustainability not only into all our operations, but also into our thinking and behaviour; it means that our handprint, the positive effect in terms of sustainability, is bigger than our footprint"
Corporation#6
1. Personal Mastery. "Our reporting reflects the view that all of our operations—and our dialogue with stakeholders—must be based on ethical and sustainable business practices, since these provide the basis for our long-term competitiveness"
2. Mental Models. "The corporate responsibility principles cover all aspects of our operations and strategy and are also integrated into the way that we conduct our business"

(continued)

Table A.2 (continued)

3. Shared Vision. "We use reporting as an opportunity to illustrate what we have to ensure that the business operations are sustainable, and to indicate actions we expect to take in the future to enhance individual well-being and the natural environment"
4. Team Learning. "Our aims are to communicate openly and transparently"
5. Systems Thinking. "Our aim of the new sustainability Key Performance Indicators (KPIs) is to secure the continual improvement and more frequent systematic monitoring of the progress. In addition, the KPIs also indicate financial savings against a target baseline, making the economic benefits of sustainability more concrete"
Corporation#7
1. Personal Mastery. "The principal channel for investors' information is our website. We have continuously developed the content of this website to improve both site content and user-friendliness"
2. Mental Models. "We have particularly focused on developing the information content and presentation of our interim reports in order to provide information to investors"
3. Shared Vision. "We have carried out an online questionnaire about investor relations conduct for key owners, investors and analysts about their expectations vis-à-vis responsibility, responsibility communication and reporting"
4. Team Learning. "We arrange frequently investors' and owners' meetings, annual general meeting and capital markets days and events intended for investors. We are active in contact with analysts. Our Financial statements, interim reports and annual report present result for investors"
5. Systems Thinking. "The corporation is included in the Forum Ethics Excellence Investment Register. Selection for inclusion in the register indicates that the corporation performs better than average in its sector in terms of corporate social responsibility"

Table A.3 Major themes raised in the discourses of the Cases of the Finnish corporations about the "corporate sustainability" in relation to 'Customers and consumers: A five disciplines classification'

Corporation#1
1. Personal Mastery. "Audits are prepared and carried out annually for each of the purchasing categories. The aim is to ensure that the Supplier Code of Conduct is followed in practice throughout the supply chain"
2. Mental Models. "Transparency and responsibility of the entire supply chain are of increasing interest to our customers and other stakeholders"
3. Shared Vision. "Our products enable our customers to make sustainable choices and improve their environmental footprint. The requirement for transparency and responsibility of the supply chain comes also from our customers and other stakeholders"
4. Team Learning. "Together with our partners, we secure a sustainable supply of raw materials for our units and a supply of renewable products for our customers"
5. Systems Thinking. "The Life Cycle Assessment (LCA) assesses the environmental impacts associated with all stages of a product's life cycle"
Corporation#2
1. Personal Mastery. "We conduct LCI surveys on all of our main products, using calculations that are updated annually. We evaluate and monitor our suppliers via self-assessments and supplier audits"

(continued)

Table A.3 (continued)

2. Mental Models. "Our Supplier Code of Conduct applies to all of our procurement operations around the world"
3. Shared Vision. "We require our suppliers and partners to comply with our sustainability requirements, including safety aspects"
4. Team Learning. "We engage with our suppliers through our supply chain approach. Our life cycle analysis experts compile life cycle inventory (LCI) data on our products. We also receive indirect stakeholder feedback through grievance channels, contacts with trade unions, and various surveys and studies including surveys of customer satisfaction and employee satisfaction. We have been devising a new e-learning tool, and group training sessions were also further improved"
5. Systems Thinking. "We use a wide range of tools including e.g.: (1) specific policies, guidelines and statements on Global Responsibility, 2 Group-level Global Responsibility targets and key performance indicators (KPIs), (3) management systems such as ISO and OHSAS, (4) supply chain management, (5) social and environmental impact and risk assessments, (6) sustainability due diligence for investment decisions and (7) responsibility reporting and third party assurance"
Corporation#3
1. Personal Mastery. "As an active corporate citizen, we offer expert advice to decision makers and non-governmental organizations in energy-related issues. We take into consideration the entire life cycle of its energy products"
2. Mental Models. "We base our customer relations on honesty and trust. We treat our suppliers and subcontractors fairly and equally and we choose them based on merit; and with the expectation that they will consistently comply with our requirements and with our Supplier Code of Conduct"
3. Shared Vision. "The Supplier Code of Conduct is based on the principles of the United Nations Global Compact and is divided into four sections: business practices, human rights, labour standards and the environment and customer relationship and products"
4. Team Learning. "Our supplier relationship management consists of the following: (1) Systematic supplier relationship management, (2) Category management model, and (3) Joint development projects with suppliers"
5. Systems Thinking. "Our products and customer relationship management consist of the following: (1) Products that meet customer needs, (2) Striving for safe, easy and long-term relationship, (3) Customer service development, (4) Services including web, social media and mobile, (5) Energy-conservation instructions and energy-efficiency services, (6) Origin-labelled electricity and (7) Product developments with customers"
Corporation#4
1. Personal Mastery. "As the majority of our products are classified as hazardous, ensuring that they are handled safely throughout their life cycle is extremely important"
2. Mental Models. "We always ensure that our customers have the information they need to handle our products safely and that the products comply with all national and international statutory requirements"
3. Shared Vision. "We provide sufficient and up-to-date information as an important part of customer communications. Our renewable products offer consumers and businesses a cleaner way to stay on the move and to transport goods"
4. Team Learning. "Safety data sheets and technical product information on products sold in markets can be consulted at the corporation's web site. Product labels also include information on safety-related questions. REACH requirements are taken into account in procurement and sales contracts, R&D, and risk management practices"

(continued)

Table A.3 (continued)

5. Systems Thinking. "In line with its cleaner traffic strategy, we offer to our customers a range of product solutions with a smaller environmental footprint"
Corporation#5
1. Personal Mastery. "The value chain of an equipment supplier typically includes raw material extraction, distribution, manufacturing, assembly, transportation and installation at our customer's site. We have calculated the carbon footprint of five our technologies"
2. Mental Models. "We work to guarantee the best return on the customer's investment with minimal ecological impact"
3. Shared Vision. "We have a globally integrated HESQ management system and an excellent track record for its safety performance in large customer projects. Through providing industry benchmark technologies and life cycle solutions to our customers, we consider ourselves to have a significant positive handprint towards sustainable use of resources"
4. Team Learning. "We support our customers in reducing the ecological footprint of their operations"
5. Systems Thinking. "We provide our customers with sustainable technologies and services that guarantee performance and lifelong benefits, such as the licence to operate, reduced energy and water consumption, high recovery, and minimized emissions, thus enabling the best return on the customer's investment"
Corporation#6
1. Personal Mastery. "We are committed to operating in an ethically sound manner through responsible business practices across the entire value chain"
2. Mental Models. "Sustainability goes to the heart of our mission to create advanced materials that are efficient, long-lasting and recyclable"
3. Shared Vision. "Continuous feedback and interaction with customers help us to improve our understanding of their needs, the challenges they face and the business environment our customers operate in"
4. Team Learning. "The corporation has implemented a 'Customer first' thinking in our sustainability work for a long time"
5. Systems Thinking. "Environmental product declarations are built on life-cycle data and help in calculating sustainability performance over a product's life cycle. The corporation has published Environmental Product Declaration (EPD) to demonstrate the good performance of its products from this perspective. EPDs help our customers to quantify the environmental performance of their solutions, for example when designing for green buildings or infrastructures"
Corporation#7
1. Personal Mastery. "We are revisiting and improving the efficiency of our ways of working to enable us to continue responding to our customers' needs. This is possible through a high level of expertise, motivated and committed people and good supervisory work"
2. Mental Models. "Our focus of safety improvement is on proactive work. Our target to reduce injuries is reflected in improved operational quality for our customers and forms the basis for personnel wellbeing"
3. Shared Vision. "In the focus area of energy efficiency, the objective is to support customers' business by providing energy-efficient solutions that reduce lifecycle costs and to improve the energy efficiency of our own operations"
4. Team Learning. "Our corporate responsibility taskforce coordinates corporate responsibility reporting and development. This task force is made up of representatives from the human resources, legal, marketing and communications, finance and the corporate responsibility and environmental organisations"
5. Systems Thinking. "Integrated management system covers quality, safety, energy and environmental aspects"

Table A.4 Major themes raised in the discourses of the Cases of the Finnish companies about the "corporate sustainability" in relation to 'Employees of the companies: A five disciplines classification'

Corporation#1
1. Personal Mastery. "We offer our employees the possibility to enhance their competences through training, internal job rotation and other learning opportunities"
2. Mental Models. "Our unified model includes early support, an assessment of work capacity and a personal work capacity development plan to ensure well-being for everyone every day"
3. Shared Vision. "Safety training is systematic and consistent, in fact we aim to keep it customised as our employees are the best source of new ideas on how to develop the safety culture in their own operational environment"
4. Team Learning. "We constantly educate our employees and management to create a healthy working environment for everyone. High environmental standards and safety performance are integral elements of our management system and are continuously improved"
5. Systems Thinking. "Preventive work is the most essential tool with which to secure a safe working environment"
Corporation#2
1. Personal Mastery. "We encourage our OHS networks to share best practices, learn from any incidents, clarify modes of operation, roles and responsibilities, and improve reporting practices"
2. Mental Models. "All new suppliers need to perform self-assessments that include questions relating to human rights, ethical business practices, environmental performance, labour practices and occupational health and safety matters"
3. Shared Vision. "Investing in our people is our strategic priority, whether this involves ethical leadership, safety issues, or helping our former colleagues to find new opportunities"
4. Team Learning. "A new IT-tool is implemented to help us to identify, categorise, assess and monitor our suppliers, and to address risks related to possible serious issues along our supply chain such as child labour, human trafficking, forced and compulsory labour, corruption and anti-trust violations"
5. Systems Thinking. "Visible leadership in safety, personal accountability and contractors' safety management will drive performance improvements. We apply our safety toolbox in all units. This toolbox includes practical, hands-on tools that directly address individuals' behaviour"
Corporation#3
1. Personal Mastery. "A key occupational sustainable performance indicator includes lost workday incident frequency for own employees which is reported monthly"
2. Mental Models. "The goals are to promote health and sustainability, support the employees' capacity to work throughout their career and promote the functionality of work communities"
3. Shared Vision. "Our aims are to have engaged and satisfied employees. We want to create attractive career and development opportunities for individuals to continuously grow their professional skills and know-how. Our EHS policies comprehensively cover issues related to employee well-being"
4. Team Learning. "Each member of the work community is responsible for their own well-being and competence and for the mutual development of well-being"
5. Systems Thinking. "Our sustainable management emphasizes the company's strategic intent to create a safe workplace. Sustainable management is based on our policies and commitments as well as instructions and company-defined minimum requirements for EHS work"

(continued)

Table A.4 (continued)

Corporation#4

1. Personal Mastery. "Process safety is based on identifying process-related risks in advance and preventing accidents. Performance in this area is regularly reviewed using internal audits and official inspections. Technical safety systems and procedures, up-to-date protective equipment, and access to the appropriate safety data sheets are all used to ensure a high standard of health and safety"

2. Mental Models. "Developing strategic competencies, managerial capabilities, and the expertise and skills of the personnel are based on helping to support the company's businesses to achieve their short- and long-term goals"

3. Shared Vision. "We promote a risk-aware culture in all areas of the company's decision-making. Our personnel receive constant training on process safety-related matters"

4. Team Learning. "We have a contractor safety development program designed to help enhance the safety of contractor work at the company's sites, improve collaboration, and develop monitoring processes, both when selecting contractors and during their on-site work"

5. Systems Thinking. "Good process safety ensures that a plant's processes operate without incident and prevents personnel from being exposed to danger and the environment from being polluted"

Corporation#5

1. Personal Mastery. "We have received a company-integrated EHSQ management certification for our global locations"

2. Mental Models. "Improving the sustainability of our own operations, supply chain management, and technology and plant safety management have been identified as important targets to meet the expectations of customers, employees and other stakeholders"

3. Shared Vision. "Continuous development of the EHSQ management systems ranks high on the company's sustainability agenda"

4. Team Learning. "The engagement includes work satisfaction, commitment, pride, loyalty, a strong sense of personal responsibility, and a willingness to be an advocate for the organization. Another key element is performance enablement, which focuses on customer service and quality, involvement, training and teamwork"

5. Systems Thinking. "We have harmonized our quality management systems and sustainability reporting to improve our performance. We have a Product Compliance management process to ensure that the plants and products engineered and delivered are reliable and meet all applied safety standards during all phases of the product life cycle"

Corporation#6

1. Personal Mastery. "Our firm objective is to minimize the environmental burden of the Group's operations as much as is economically and technically feasible. The basis of this work is EHSQ Policy"

2. Mental Models. "The ultimate goal remains zero accidents and ambition levels have been set to make a step change improvements"

3. Shared Vision. "We are committed to providing a safe and healthy working environment at its production sites and facilities for our own personnel, contractors and visitors"

4. Team Learning. "The number of non-conformities and corrective actions in EHSQ and energy management systems found by external auditors in our units are regularly monitored"

5. Systems Thinking. "Operational efficiency of our EHSQ and energy management systems and certification is monitored using both internal and external audits and ensured by co-operating with certification bodies. Our production sites employ Environmental Management Systems (EMS) and risk-based management systems"

(continued)

Table A.4 (continued)

Corporation#7
1. Personal Mastery. "The employees of the company and the contractors are required to have completed occupational safety card training"
2. Mental Models. "The objective is to reduce the total number of injuries, to increase awareness to eliminate human errors and to create a safety culture"
3. Shared Vision. "Safety is promoted by ensuring that subcontractors train their own people in safety matters. New service providers are required to have undergone safety training before they commence work at a plant"
4. Team Learning. "We perform supplier evaluations of new potential suppliers and of existing suppliers who are deemed critical and whose performance requires improvement"
5. Systems Thinking. "We continuously evaluate our own operations and those of our partners to ensure compliance with the highest business standards, safety, quality and environmental matters. Principal subcontractors are invited once a year to a joint safety briefing, which covers the latest safety topics, gives feedback on the development of supplier safety and states the targets"

Table A.5 Major themes raised in the discourses of the Cases of the Finnish companies in the way they apply the "corporate sustainability" to 'Communities and authorities: A five disciplines classification'

Corporation#1
1. Personal Mastery. "Our contribution to the community's well-being derives from direct and indirect employment and paying taxes, among others. We generate well-being at work, in local communities and throughout society at large, and are committed to global sustainability principles"
2. Mental Models. "In producing electricity and heat to our factory, the plant will also provide renewable energy for the surrounding community"
3. Shared Vision. "One of our most fundamental duties is to guarantee that all our operations are sustainable and all our products are safe for both people and the environment. In addition to producing heat to the factory, it will provide renewable energy for the surrounding community.
4. Team Learning. "By behaving responsibly towards our employees and society, we can improve the quality of life of our stakeholders"
5. Systems Thinking. "We focus on the following: (1) Participation in cooperation with governance bodies, (2) Society, local communities and NGOs Cooperation projects, (3) Image and brand surveys, (4) Site visits, hearings, meetings and interviews, (5) Cooperation with legislators and decision makers, industry and trade associations, (6) Participation in industry and trade associations' work, (7) Media Meetings, interviews, press events and site visits, and (8) Academics Joint research programmes with universities and research institutes"
Corporation#2
1. Personal Mastery. "By creating shared value, we aim to increase our competitiveness, while at the same time promoting improvements on economic, environmental, ethical, and social issues along our value chain"
2. Mental Models. "We recognize the need to create shared value in the communities where we operate"
3. Shared Vision. "Our significant stakeholder groups are the following: (1) Investors, (2) Local communities, (3) Partners and suppliers, (4) Governments, (5) Consumers. (6) Customers, (7) Employees, (8) Media and (9) NGOs"

(continued)

Table A.5 (continued)

4. Team Learning. "We obtain stakeholder feedback largely through our regular contacts with our stakeholders at meetings, fairs, community visits, public hearings, open house events or other events organized by or for our stakeholders"
5. Systems Thinking. "We conduct Environmental and Social Impact Assessments (ESIA) for all new projects that could cause significant adverse impacts or other substantial changes in local conditions. The results of ESIAs give us valuable information on how local communities will be affected by changes in their socio-economic structure, impacts on cultural heritage, and developments in community health, safety and security"
Corporation#3
1. Personal Mastery. "General public's expectations of our actions in society include e.g. (1) Activities for the good of society, (2) Transparency, (3) Tax payment and development of the energy sector to meet society's needs, (4) Easily accessible through media desk and (5) Continuous development of crisis communication preparedness"
2. Mental Models. "We promote well-being and sustainability in the work community, respect for individuals and mutual trust, and responsible operations in our supply chain and in society in large"
3. Shared Vision. "The aim of our CARE programme is to promote health, sustainable, employee work capacity and work community functionality. The programme activities are tailored to take into account the legislative requirements and unique cultural aspects in different countries. We distribute energy to our customers while taking into consideration long-term, sustainable community planning"
4. Team Learning. "Our social responsibility includes being a good corporate citizen and taking care of our own personnel and the surrounding community"
5. Systems Thinking. "Authorities' and decision makers' expectations about our actions include the following: (1) Compliance with laws and regulations, (2) Payment of taxes and dividends, (3) Maintaining dialogue, (4) Transparency and reliable reporting, (5) Compliance management, (6) Publishing tax footprint, (7) Active and open dialogue with authorities and decision makers, (8) Active and open communication, reports assured by a third party, (9) Use of environmentally benign forms of energy, (10) Good employer and neighbour, (11) Meetings with local residents and customers and (12) Support for local community activities"
Corporation#4
1. Personal Mastery. "We implement open and timely communication with the local community. Our operations have a major impact on society. We provide them with the possibility to give feedback via the internet. We constantly monitor the environmental impact of our operations and ensure that we operate within the terms of our environmental permits"
2. Mental Models. "Wherever we operate, we ensure that our operations have the minimum possible negative impact on the natural environment and the surrounding community. Our investments develop e.g. sustainable production capacity, energy efficiency, local infrastructure and electricity distribution reliability"
3. Shared Vision. "We distribute newsletters to people living near to our plant. We do not make a profit at the expense of safety or security"
4. Team Learning. "We continue to constantly monitor the environmental impact of our operations and ensure that we operate within the terms of our environmental permits. We collaborate with other companies based close to our sites. We arrange open door days at our refineries. A Facebook site is maintained by our refineries. We arrange meetings with local municipal leaders"
5. Systems Thinking. "Magazines and newsletters are distributed to the surrounding community. Our focus is on the following aspects: (1) Compliance with the law and statutory regulations, (2) Sustainable operations, (3) Good risk management, (4) Reliable and sufficient reporting,

(continued)

Table A.5 (continued)

(5) A good taxpayer, (6) Good overall return on the company's share, (7) Sufficient and reliable information, (8) Effective management of the environmental impact, (9) Outreach events for people living near the factories, (10) Collaboration with other companies, (11) Support for the local community, (12) Annual General Meeting, (13) Media events and interviews, and (14) Visits to our factories"
Corporation#5
1. Personal Mastery. "Our company culture encourages everyone to discuss and develop our operations. We are in a good position to deliver solid solutions that are safe, environmentally sound and can be accepted and embraced by the local community, ensuring social licence to operate for our customers"
2. Mental Models. "Our aim is to be an open and equal work community. Through providing sustainable solutions to our customers, we create jobs and wealth locally in countries where we operate or where our customers' projects are located"
3. Shared Vision. "Our aim is to be an open and equal work community. The company culture encourages everyone to discuss and develop our operations. We want to support local projects in connection with major solution deliveries to our customers"
4. Team Learning. "Our community projects are based on the needs of the local community in the project delivery location in question, and they are defined in a dialogue with the local community"
5. Systems Thinking. "We aim at completing community projects jointly with our customers with joint financing. We have made a significant global impact by creating new revenue streams, reducing our customers' environmental footprints, and increasing well-being in local communities"
Corporation#6
1. Personal Mastery. "Our operations have economic impacts on local, national and global communities n through paying taxes, direct and indirect employment, and other means of community involvement"
2. Mental Models. "We continue to pursue transparency, learning from dialogue with customers, investors, non-governmental organizations and local communities"
3. Shared Vision. "By paying special attention to waste management and segregation techniques, many waste fractions resulting from production operations are now recycled"
4. Team Learning. "Finding balance between global market trends and responsibility towards communities is sometimes difficult, especially in economic downturns"
5. Systems Thinking. "Events tailored to creating open and active participation, such as a variety of 'open house' events on production sites, are clear signals for local communities that we want to be part of local communities and to encourage an open culture"
Corporation#7
1. Personal Mastery. "The focus areas on the environmental responsibility front are to strengthen product-related energy efficiency and lifecycle know-how, to develop energy efficient production, to increase recycling and material efficiency and to maintain responsible operations"
2. Mental Models. "Our strategy and core competence support the partnership with the community"
3. Shared Vision. "We participate in diverse ways in the activities of the neighbouring community. Discussion between our representatives and local decision-makers and regulators is an inherent part of our everyday operations"

(continued)

Table A.5 (continued)

4. Team Learning. "The idea behind the collaboration projects is to use training and information sharing to get people to work for a better environment. Examples of cooperation include work done to safeguard the working conditions and improve productivity of the shared harbour"
5. Systems Thinking. "The aim is long-term partnership with the community. Our main sponsor efforts include e.g. research, training and other longstanding cooperation with communities"

References

AA1000APS (2008) Account ability principles standard. http://www.accountability.org/standards/aa1000aps.html. Accessed 21 July 2014

BS OHSAS 18001 (2007) Occupational health and safety management systems. http://www.isoqsltd.com/iso-certification/bs-ohsas-18001/. Accessed 21 July 2014

Christensen CM (1997) The Innovator's dilemma: when new technologies cause great firms to fail. Harvard Business Review Press, Boston

Christmann P (2000) Effects of best practices of environmental management on cost advantage: the role of complementary assets. Acad Manage J 43(4):663–680

Corporation#1 (2014) Metsä Group corporation—sustainability report 2013, 67 pp. http://www.metsagroup.com/Pages/Default.aspx. Accessed 21 July 2014

Corporation#2 (2014) Stora Enso corporation—global responsibility report 2013, 80 pp. http://assets.storaenso.com/se/com/DownloadCenterDocuments/Stora_Enso_Global_Responsibility_Report_2013.pdf. Accessed 21 July 2014

Corporation#3 (2014) Fortum corporation—sustainability report 2013, 108 pp. http://fortum-ar-2013.studio.crasman.fi/pub/web/pdf/Fortum_Sustainability_2013.pdf. Accessed 21 July 2014

Corporation#4 (2014) Neste Oil corporation—sustainability report 2013, 315 pp. http://2013.nesteoil.com/sustainability/Managing-sustainability-and-strategy/. Accessed 21 July 2014

Corporation#5 (2014) Outotec corporation—sustainability report 2013, 47 pp. http://www.outotec.com/ImageVaultFiles/id_1400/cf_2/Outotec_SUSTAINABILITY_2013.PDF. Accessed 21 July 2014

Corporation # 6 (2014) Outokumpu corporation—sustainability report 2013, 37 pp. http://www.outokumpu.com/SiteCollectionDocuments/Sustainability-2013.pdf. Accessed 21 July 2014

Corporation#7 (2014) Rautaruukki corporation—corporate responsibility report 2013, 81 pp. http://www.ruukki.com/~/media/Files/Corporate%20responsibility/CR%20report%202013/Ruukki_Corporate_Responsibility_report_2013.ashx. Accessed 21 July 2014

Cunliffe A (2008) Orientations to social constructionism: relationally responsive social constructionism and its implications for knowledge and Learning. Manag Learn 39(2):123–139

de Lange DE, Busch T, Delgado-Ceballos J (2012) Sustaining sustainability in organizations. J Bus Ethics 110(2):157–172

Denzin NK, Lincoln YS (1998) Collecting and interpreting qualitative materials. Sage Publications, London

Donaldson T, Preston L (1995) The stakeholder theory of the modern corporation: concepts, evidence and implications. Acad Manage Rev 20(1):65–91

Drucker PF (2001) The essential drucker: management challenges for the 21st century. Buttenvorth-Heinemann Publications, London

Du S, Bhattacharya CB, Sen S (2010) Maximizing business returns to corporate social responsibility. Bus Ethics Eur Rev J Manag Rev 12(1):8–19

Dyllick T, Hockerts K (2002) Beyond the business case for corporate sustainability. Bus Strategy Environ 11(2):130–141. (Special Issue: Sustainability at the Millennium: Globalization, Competitiveness and Public Trust)

Eriksson P, Kovalainen A (2008) Qualitative methods in business research. Sage Publications, London
Fairhurst G, Grant D (2010) The social construction of leadership: a sailing guide. Manag Commun Q 24(2):171–210
Fairclough N (1996) Discourse and social change. Polity Press, Cambridge
Fairclough N (2005) Critical discourse analysis, organizational discourse, and organizational change. Organ Stud 26(2):915–939
Fisher C, Lovell A (2006) Business ethics and values: individual, corporate and international perspectives, 2nd edn. Pearson Education Limited, Harlow
Flick U (2006) An introduction to qualitative research, 5th edn. Sage Publications, London
Fombrun CJ (2005) Corporate reputations as economic assets. In: Hitt MA, Freeman RE, Harrison JS (eds) Handbook of strategic management. Blackwell Publishing, Oxford, pp 289–312
Freeman RE (1984) Strategic management: a stakeholder approach. Pitman, Boston
Freeman RE (1994) The politics of stakeholder theory: some future directions'. Bus Ethics Q 4 (4):409–421
Garriga E, Melé D (2004) Corporate social responsibility theories: mapping the territory. J Bus Ethics 53(1–2):51–71
GRI (2014) Global reporting initiatives: sustainability reporting guidelines. https://www.globalreporting.org/reporting/g4/Pages/default.aspx. Accessed 21 July 2014
Greimas AJ (1987) On meaning: selected writings in semiotic theory. University of Minnesota Press, Minneapolis
Greimas AJ (1990) The social sciences: a semiotic view. University of Minnesota Press, Minneapolis
Greimas AJ, Courtes J (1982) Semiotics and language: an analytical dictionary. Indiana UP, Bloomington
Guba EG, Lincoln YS (1989) Fourth generation evaluation. Sage Publications, Newbury Park
Hannah ST, Avolio BJ, Walumbwa FO (2011) Relationships between authentic leadership, Moral courage, and ethical and pro-social behaviours. Bus Ethics Q 21(4):555–578
Hébert L (2011) Tools for text and image analysis: an introduction to applied semiotics. Université du Québec à Rimouski, Québec. http://www.signosemio.com/documents/Louis-Hebert-Tools-for-Texts-and-Images.pdf. Accessed 21 July 2014
ILO (2014) Rights at work. ILO. http://www.ilo.org/global/about-the-ilo/decent-work-agenda/rights-at-work/lang–en/index.htm. Accessed 21 July 2014
ISO 9001 (2008) Quality management systems. International Organization for Standardization http://www.isoqsltd.com/iso-certification/iso-9001/. Accessed 21 July 2014
ISO 14001 (2004) Environmental management systems. International Organization for Standardization. http://www.isoqsltd.com/iso-certification/iso-14001/. Accessed 21 July 2014
Jackson MC (2003) Systems thinking: creative holism for managers. Wiley, West Sussex. http://webcourses.ir/dl/Systems%20Thinking."pdf. Accessed 21 July 2014
Jones T (1995) Instrumental stakeholder theory: a synthesis of ethics and economics. Acad Manag Rev 20(2):404–437
Kujala J (2010) Corporate responsibility perceptions in change: finnish managers' views on stakeholder issues from 1994 to 2004. Bus Ethics Eur Rev 10(3):233–247
Kvale S, Brinkmann S (2009) Interviews: learning the craft of qualitative research interviewing, 2nd edn. Sage Publications, London
Leeman JE (2002) Applying interactive planning at DuPont. Syst Pract Action Res 15(2):85–109
Lincoln YS, Guba E (1985) Naturalistic inquiry. Sage Publications, Beverly Hills
Lord RG, Dinh JE (2012) Aggregation processes and levels of analysis as organizing structures for leadership theory. In: Day DV, Antonakis J (eds) The nature of leadership, 2nd edn. Sage Publications Inc., Thousand Oaks, pp 29–65
Maon F, Lindgreen A, Swaen V (2010) Organizational stages and cultural phases: a critical review and a consolidative model of corporate social responsibility development. Int J Manage Rev 12 (1):20–38

OECD (2011) OECD guidelines for multinational enterprises. OECD Publishing. http://dx.doi.org/10.1787/9789264115415-en. Accessed 21 July 2014

Piccolo RF, Greenbaum R, Den Hartog DN, Folger R (2010) The relationship between ethical leadership and core job characteristics. J Organ Behav 31(2–3):259–278. (Special Issue: Putting Job Design in Context)

Potter J (1997) Discourse analysis as a way of analysing naturally occurring talk. In: Silverman D (ed) Qualitative research: theory, method and practice. Sage Publications, London, pp 144–160

Potter J (2000) Post cognitivist psychology. Theor Psychol 10(2):31–37

Salzmann O, Steger U, Ionescu-Somers A, Baptist F (2006) Inside the mind of stakeholders—are they driving corporate sustainability? International Institute for Management Development Global Corporate Governance Research Initiative. IMD Working Paper, Oct 2006

Schein EH (1996) Leadership and organizational culture. In: Hesselbein F, Goldsmith M, Beckhard R (eds) The leader of the future: new visions, strategies, and practices for the next era. Jossey-Bass Inc., San Francisco, pp 59–70

Senge PM (1994) The fifth discipline: the art and practice of the learning organization, 2nd edn. Bantam Doubleday/Currency Publishing Group, New York

Uhl-Bien M, Maslyn J, Ospina S (2012) The nature of relational leadership: a multitheoretical lens on leadership relationships and processes. In: Day DV, Antonakis J (eds) The nature of leadership, 2nd edn. Sage Publications Inc., Thousand Oaks, pp 289–330

UNCED (1992) United Nations conference on environment and development. Rio de Janeiro. United Nations Department of Economic and Social Affairs (DESA), New York

United Nations Global Compact (2014) The ten principles. http://www.unglobalcompact.org/. Accessed 21 July 2014

van Dijk TA (1995) Discourse semantics and ideology. Discourse Soc 6(2):243–289

van Dijk TA (2001) Critical discourse analysis. In: Schiffrin D, Tannen D, Hamilton HE (eds) The handbook of discourse analysis. Blackwell Publishing, Massachussets and Oxford, pp 363–415

van Marrewijk M (2003) Concepts and definitions of CSR and corporate sustainability: between agency and communion. J Bus Ethics 44(2–3):95–105

Väyrynen S, Koivupalo M, Latva-Ranta J (2012) A 15 year development path of actions towards an integrated management system: description, evaluation and safety effects within the process industry network in Finland. Int J Strateg Eng Asset Manag 1(1):3–32

Väyrynen S, Jounila H, Latva-Ranta J (2014) HSEQ assessment procedure for supplying industrial network: a tool for implementing sustainability and responsible work systems into SMES. In: Ahram T, Karwowsk W, Marek T (eds) Proceedings of the 5th International conference on applied human factors and ergonomics AHFE 2014, Kraków, Poland, 19–23 July 2014

Wagner M (2005) How to reconcile environmental and economic performance to improve corporate sustainability: corporate environmental strategies in the European paper industry. J Environ Manage 76(2):105–118

Webley S, Werner A (2008) Corporate codes of ethics: necessary but not sufficient'. Bus Ethics Eur Rev 17(4):405–415

Weick KE (2002) Making sense of the organization. Blackwell, Oxford

Wetherell M (1998) Positioning and interpretative repertoires: conversation analysis and post-structuralism in dialogue. Discourse Soc 9(1):387–412

Wetherell M, Taylor S, Yates SJ (2001) Discourse theory and practice: a reader. Sage, London

Wheeler D, Colbert B, Freeman RE (2003) Focusing on value: reconciling corporate social responsibility, sustainability and a stakeholder approach in a network world. J Gen Manag 28(2):1–28

Windell K (2006) Corporate Social Responsibility under Construction: Ideas, Translations, and Institutional Change. Upsala Universitet. Företagsekonomiska institutionen Department of Business Studies, Doctoral Thesis No. 123. Universitettryckeriet, Ekonomikum, Uppsala

Wood DJ (1991) Corporate social performance revisited. Acad Manage Rev 16(4):691–718

Wood DJ, Jones RE (1995) Stakeholder mismatching: a theoretical problem in empirical research on corporate social performance. Int J Organ Anal 3(3):229–267

Work JW (1996) Leading a diverse work force. In: Hesselbein F, Goldsmith M, Beckhard R (eds) The leader of the future: new visions, strategies, and practices for the next era. Jossey-Bass Inc., San Francisco, pp 71–80

World Business Council for The "corporate sustainability" (2000) Corporate social responsibility: making good business sense. World Business Council for The "corporate sustainability", Geneve

WCED (1987) Our common future. World Commission on Environment and Development. Oxford University Press, Oxford

Yin R (1994) Case study research: design and methods, 2nd edn. Sage Publications, Thousand Oaks

Yukl G (1998) The nature of leadership, 4th edn. Prentice-Hall International Inc., London

Zink KJ (2014) Designing sustainable work systems: the need for a systems approach. Appl Ergon 45(1):126–132. (Special Issue: Systems Ergonomics/Human Factors)

Part IV
Conclusions

Chapter 15
Conclusions

Seppo Väyrynen, Kari Häkkinen and Toivo Niskanen

This book proposes a movement towards collaborative and shared leadership approaches. It views OSH and performance excellence as entailing a broader examination of leadership relationships and practices. The complexity of leadership is explored through the impact that contexts (e.g. national and organizational culture) may have on leaders. Different views on bottom-up and top-down processes are integrated.

This book articulates in each chapter the linkages with literature, then proceeds to show a more comprehensive practical approach. It proposes that greater clarity in understanding leadership in OSH and performance excellence can be developed by addressing two fundamental issues. First, how do subunit inputs and processes combine to produce unit-level outcomes and how does leadership affect this process? Second, how do leaders influence the way that individual- and team-level inputs are combined to produce organizational outputs?

To reiterate the main ideas stated in the preface, we conclude that generally, this book could follow, utilize and contribute practically in terms of all the 15 key starting points, as presented in some chapters. Nonetheless, most of the chapters focus mainly on one or some of the points, as follows:

S. Väyrynen (✉)
Faculty of Technology, Industrial Engineering and Management, University of Oulu,
P.O. Box 4610, FI - 90014 Oulu, Finland
e-mail: Seppo.Vayrynen@oulu.fi

K. Häkkinen
If P&C Insurance Company Ltd, Niittyportti 4, P.O. Box 1032, 00025 IF Espoo, Finland
e-mail: kari.hakkinen@if.fi

T. Niskanen
Ministry of Social Affairs and Health, Occupational Safety and Health Department,
Legal Unit, Box 33, 00023 Government Helsinki, Finland
e-mail: toivo.niskanen@stm.fi

1. Comprehensive and widening OSH both as a specific process and embedded in the work system and processes
2. Total quality by participation
3. Excellence in leadership and management
4. Regulation-based and voluntary
5. Effectiveness of policies
6. Insurance against losses
7. OSH culture
8. European Union approach
9. Responsibility and well-being embedded in holistic sustainability
10. "Macro" stakeholders: the state, social partners and closer operational ecosystem of partners
11. Integrated HSEQ management
12. Integrated management, not only intra-organizational
13. Utilization of research contributions
14. Harm at work and at leisure
15. ICT—more important enabler

Now more often perceived as an essential part of an integrated and multiple, comprehensive management system, OSHM is a complicated puzzle of many pieces. As editors and authors, we contribute to solving this puzzle with this book, *Integrated Occupational Safety and Health Management: Solutions and Industrial Cases*. We hope that the forthcoming general, international standardized guidelines for OSHM under ISO/CD 45001 will be one step forward to make this issue more well known in the extensive field of management and quality. This book is expected to add practical value and tools to ISO/CD 45001 that is described as follows:

> The overall aim of the standard remains the same and those familiar with OHSAS 18001 will recognize many of the themes in the new ISO standard. However, there have been some very interesting developments related to the new rules for developing International Management System Standards. For example, there is now a much stronger focus on the "context" of an organization as well as a stronger role for top management and leadership. In the new standard, an organization has to look beyond its immediate health and safety issues and take into account what the wider society expects of it. Organizations have to think about their contractors and suppliers as well as, for example, how their work might affect their neighbours in the surrounding area. This is much wider than just focussing on the conditions for internal employees and means organizations cannot just contract out risk.

The top-management teams need to ensure that the organisation in which the resources will be working is itself capable of succeeding—and in making this assessment, managers must scrutinize whether the organization's processes and values fit the OSH problems with HSEQ systems. On the other hand, if the ways of getting work done and decision making in the business of corporate sustainability would impede rather than facilitate the work of the new teams—because different people need to interact with various parties about diverse subjects and with different timing than has habitually been necessary—then a new team structure may be necessary to integrate OSH and HSE matters. Additionally, one of the dilemmas is that by their very nature, the processes are established so that managers and employees perform recurrent tasks in collaboration and in a consistent way, time after time through tightly controlled procedures. However, the clear, consistent and effective best practices also define what an organization can do. In this context, a company's values, by necessity, must reflect its management practices and business model because these define the practices that its managers must follow to implement the effective, proactive measures in business and corporate sustainability. Some processes are formal, in the sense that they are explicitly defined, visibly documented and consciously followed. Other processes are informal, which people follow simply because they work or because "that's the way we do things around here". The organizational culture is a powerful management tool in promoting best practices. Furthermore, excellence in organizational culture enables the managers and employees to act cooperatively.

Chapter 2 has discussed the developments in understanding safety management. There seems to be ample evidence of the elements that are related to satisfactory safety performance. These include good overall management, safety goals and competence requirements defined for managers, visibility and commitment by top management, rewards and incentives for safe work, effective learning from accidents and incidents and continuous improvement using risk assessments, effective safety inspections and internal audits, as well as participative leadership practices, such as good communication and trust. Safety management systems and technical standards for safe work environments have been established in highly developed economies, thus, implying the formal safety approach. In the future, there seems to be a need for better handling of the more informal processes in safety, that is, human behaviour, attitudes and safety culture. Furthermore, the complexity of health and safety as a leadership mission calls for the improvement of the general management theory to deal better with the complex, human-centred issues in management.

Chapters 3 and 4 have demonstrated the accident risks and challenges in safety management in two occupational areas where accidents happen more often than in most other occupations: construction work (Chap. 3) and industrial maintenance (Chap. 4). In the construction industry case, the improvements seem to be strongly related to the sharpened role of top management. As a consequence of increased safety focus from the top, supervisors and employees alike also pay more attention to safety. The way towards a zero-accident culture in the construction industry entails continual and persistent efforts through which significant improvements in the results may occur. The risk profiles of both construction and maintenance work seem to have partly similar patterns. Increased focus is needed for the planning of the safety aspects of the individual working tasks, as well as cooperation with customers and subcontractors in safety issues.

Chapter 5 has described the different methods used for supplier companies' HSEQ management. The principal company has selected the main methods: HSEQ AP and after-contract HSEQ evaluation, in addition to general occupational safety forms for suppliers. Supplier companies' experiences of HSEQ AP have been encouraging but further development is needed. It could be clarified that safety indicators (especially, the lost time injury frequency rate, LTIFR) and the results of HSEQ AP have been improved in both principal and supplier companies, but at the same time, many other safety activities have been involved as well. Thus, it is not clear at the end how much HSEQ AP has affected this positive development. Moreover, research for additional safety indicators is needed to prove more specifically the actual safety performance level, especially for supplier companies.

Chapter 6 has primarily addressed the empirical interview segment of the whole study, conducted mainly in the seaports in Northern Finland. Some information on HSEQ was also gathered during a visit to Russia's Murmansk Commercial Seaport, in the largest part of the Barents Region. The latter part consisted of a thorough review of scientific and industrial documents; the former part referred to hearing about HSEQ AP and HSEQ issues, generally in the field, from the interviewed representatives of a maritime company network (a case). The combination of these approaches has provided recommendations and harmonizing scenarios for further consideration, as far as the wide company network of contemporary seaports is concerned.

Chapter 7 has pointed out that even though new technologies have and will emerge to ease drivers' work, the job performed in environments other than truck cabs still involves tasks that require physical activities and pose risks of occupational diseases and accidents. Thus, drivers' safety at work and work ability issues remain an area that needs continuous, systemic development. The results have inevitably showed that to successfully improve L/SH drivers' work, the relevant stakeholders' participation and a systemic approach are crucial.

Content analysis, the most essential method of Chap. 8, has been applied to identify the accident factors from the investigation reports of fatal accidents. The accident factors have been divided into two main classes, in which the factors related to safety culture and safety attitudes have been discussed. The accident factors have also been considered according to the five elements of the work system model. Finally, as in the Finnish investigation procedure, the prevention measures of accidents have been presented in a condensed form.

Based on a new kind of comprehensive data on all injurious accidents (10-year timespan) involving employees ($n = 13{,}000$) of two large workplaces in Finland, Chap. 9 has revealed the total risk according to both frequency and absence times related to all accidents. The accident categories have covered accidents at work, at home and during leisure time, as well as when commuting to or from the job site. This study's main aim has been to clarify and assess the significance of risks among employees in different LTI cases. Therefore, as far as the studied cases (a metal processing mill and a municipality) are concerned, it has been possible to show that home and leisure-time accidents appear to be the most numerous category. Preliminary analysis of the metal processing case seems to indicate that blue-collar employees have more numerous accidents at leisure and in the workplace, compared with white-collar employees.

Chapter 10 has dealt with a new kind of safety training in construction, where high accident statistics are so common in most companies. The HSEQ Training Park involves a novel safety training programme that enables practical demonstrations and active participation. The park was constructed in Northern Finland through the cooperation of almost 70 companies and communities. It began its activities with trainer trainings in the spring of 2014. This study has provided an overview of the design, construction process and structure of the training park.

Chapter 11 has addressed proactive OSH measures for the safe use of chemicals and risk prevention. The risk assessment should not be a one-off activity but ongoing as part of the process of continuous improvement. Information on the results of risk assessment should be communicated to all interested partners, for example, as an agenda item at all staff meetings. In OSH management, certain measures should be applied proactively, for example: (1) in the implementation of risk assessment in planning and leading the work, (2) in the development of supervisors' leadership skills and (3) in the follow-up of improvements in working conditions. The management should make every effort to ensure that workers have the opportunity to participate in the prevention of safety-related pitfalls, such as problems discovered in risk assessment.

Chapter 12 has highlighted the management, collaboration and safe processes in OSH. Satisfactory safety performance depends on the commitment of the top management and the collaboration of employees. In dynamic processes such as those used by the chemical industry, legislation based on a command-and-control approach is inadequate; instead, a fundamentally different view of work system procedures (e.g. self-regulation) is required. An organization needs an informed culture in which those who manage and operate the system have up-to-date knowledge about the human, technical, organizational and environmental factors that determine the organization's OSH as a whole. Rather than striving to control OSH behaviour by simply complying with OSH legislation, the focus should be on proactive activities and encouraging collaboration in OSH. Moreover, a safety culture is created as a product of the systemic structure built on the foundations of the joint actions of different parties in the management, collaboration and OHC systems.

Chapter 13 has suggested that key stakeholders' favourable perceptions of companies be crystallized into tangible assets and that the CSR reputations have intrinsic economic values because they affect a company's bottom-line performance. Financial markets now demand that companies ensure transparency in their operations; large financial institutions can exert a major influence on how companies operate in society. The CSR is perceived as the aspect of an organization's overall strategic management that focuses specifically on the social and ethical issues embedded into the functioning and decision processes of the firm. The CSR processes occur in the context of both organizational (internal) and environmental (external) forces. By engaging in CSR activities, companies can not only generate favourable stakeholder attitudes and better support behaviours, but also build a solid corporate image, strengthen stakeholder-company relationships and enhance stakeholders' advocacy behaviours over the long term.

Chapter 14 has proposed that corporate sustainability be applied as a mechanism for transferring resources and capabilities across borders of the strategic management and HSEQ management to be efficient for its purpose. Some corporate values are ethical in nature, such as those that guide the decisions and especially the processes. Successful companies' managers and employees gradually come to realize that corporate sustainability's priorities in operations and decision-making methods are the right way to work. For example, systematic leadership towards corporate sustainability could imply utilization of systems thinking in order to take step-wise approaches to transformative changes. Management quality is a key factor because it will determine a business model's success, including corporate capabilities to acquire, combine and utilize valuable resources in ways that deliver value to stakeholders.